Lecture Notes
in Business Information Processing
449

More information about this series at https://link.springer.com/bookseries/7911

Mohamed Fakir · Mohamed Baslam ·
Rachid El Ayachi (Eds.)

Business
Intelligence

7th International Conference, CBI 2022
Khouribga, Morocco, May 26–28, 2022
Proceedings

Editors
Mohamed Fakir ⓘ
Sultan Moulay Slimane University
Beni-Mellal, Morocco

Mohamed Baslam ⓘ
Sultan Moulay Slimane University
Beni Mellal, Morocco

Rachid El Ayachi ⓘ
Sultan Moulay Slimane University
Beni-Mellal, Morocco

ISSN 1865-1348   ISSN 1865-1356 (electronic)
Lecture Notes in Business Information Processing
ISBN 978-3-031-06457-9   ISBN 978-3-031-06458-6 (eBook)
https://doi.org/10.1007/978-3-031-06458-6

This Springer imprint is published by the registered company Springer Nature Switzerland AG
The registered company address is: Gewerbestrasse 11, 6330 Cham, Switzerland

Preface

In 2022 we had the great opportunity to organize the 7th edition of the International Conference on Business Intelligence (CBI 2022). The conference has grown to be a reputable event for the scientific and business communities. This book of proceedings collects papers accepted for presentation at CBI 2022.

CBI 2022 was organized by the Faculty of Sciences and Techniques (FST) and the laboratory of Information Processing and Decision Support (TIAD) at Sultan Moulay Slimane University along with the Association of Business Intelligence (AMID), and held during May 26–28, 2022, in Khouribga, Morocco.

For this edition, we received 68 submissions, which were reviewed by a Program Committee of 72 international experts in various fields related to business intelligence and decision support. Out of these submissions, the Program Committee decided to accept 23 regular papers, yielding an acceptance rate of 33.8%. The contributions are organized in topical sections: Decision Support and Artificial Intelligence; Business Intelligence and Database; and Optimization and Dynamic Programming.

As program chairs of CBI 2022 and editors of these proceedings, we would like to thank the President of Sultan Moulay Slimane University and the Dean of the Faculty of Sciences and Techniques for their support to the conference. In addition, we want to warmly thank again all the authors for their high-quality contributions and all the Program Committee members for their invaluable hard work. We also sincerely thank our keynote speakers for sharing their precious insights and expertise. Finally, our special thanks go to the Organizing Committee and to all the local arrangements coordinators. We cordially invite you to visit the CBI website at https://www.cbi-bm.com/ and to join us at future CBI conferences.

April 2022

Mohamed Fakir
Mohamed Baslam
Rachid El Ayachi

Organization

General Chair

Mohamed Fakir   Sultan Moulay Slimane University, Morocco

Program Chairs

Mohamed Baslam	Sultan Moulay Slimane University, Morocco
Rachid El Ayachi	Sultan Moulay Slimane University, Morocco

Steering Committee

A. Manuel de Oliveira Duarte	University of Aveiro, Portugal
Abderrahim Salhi	Sultan Moulay Slimane University, Morocco
Belaid Bouikhalene	Sultan Moulay Slimane University, Morocco
Brahim Minaoui	Sultan Moulay Slimane University, Morocco
Charki Daoui	Sultan Moulay Slimane University, Morocco
Driss Ait Omar	Sultan Moulay Slimane University, Morocco
Ebad Banissi	London South Bank University, UK
Halima El Biaze	University of Quebec at Montreal, Canada
Hammou Fadili	National Conservatory of Arts and Crafts, France
Hicham Zougagh	Sultan Moulay Slimane University, Morocco
Karim Benouaret	Université Claude Bernard Lyon 1, France
Majed Haddad	University of Avignon, France
Mohamed Biniz	Sultan Moulay Slimane University, Morocco
Mohamed Erritali	Sultan Moulay Slimane University, Morocco
Mostafa Jourhmane	Sultan Moulay Slimane University, Morocco
Mourad Nachaoui	Sultan Moulay Slimane University, Morocco
Muhammad Sarfraz	Kuwait University, Kuwait
Najlae Idrissi	Sultan Moulay Slimane University, Morocco
Nora Tigziri	Université Mouloud Mammeri de Tizi-Ouzou, Algeria
Suliman Hawamdeh	University of North Texas, USA
Youssef El Mourabit	Sultan Moulay Slimane University, Morocco

Program Committee

A. Manuel de Oliveira Duarte	University of Aveiro, Portugal
Abdelali Elmoufidi	Sultan Moulay Slimane University, Morocco
Abdelhak Mahmoudi	Mohammed V University, Morocco
Abdelkaher Ait Abdelouahad	Chouaib Doukkali University, Morocco
Abdellatif Dahmouni	Chouaib Doukkali University, Morocco
Abderrahim Salhi	Sultan Moulay Slimane University, Morocco
Abderrazak Farchane	Sultan Moulay Slimane University, Morocco
Abdullah Saif Altobi	Sultan Qaboos University, Oman
Adnan Mahmood	Macquarie University, Australia
Ali Kartit	Chouaib Doukkali University, Morocco
Antonio Lucadamo	University of Sannio, Italy
Aroa Casado	University of Barcelona, Spain
Ashesh Mahidadia	University of New South Wales, Australia
Belaid Bouikhalene	Sultan Moulay Slimane University, Morocco
Bernhard Bauer	University of Augsburg, Germany
Charki Daoui	Sultan Moulay Slimane University, Morocco
Chih-Ming Chu	National Dong Hwa University, Taiwan
Daniel O'Leary	University of Southern California, USA
Driss Ait Omar	Sultan Moulay Slimane University, Morocco
Farzaneh Elahifasaee	Shanghai Jiao Tong University, China
Goce Trajcevski	Northwestern University, USA
Hajar El Hammouti	King Abdullah University of Science and Technology, Saudi Arabia
Halima Elbiaze	Université du Québec à Montréal, Canada
Hamid Aksasse	Ibn Zohr University, Morocco
Hamid Garmani	Sultan Moulay Slimane University, Morocco
Hammou Fadili	National Conservatory of Arts and Crafts, France
Han-Chieh Chao	National Dong Hwa University, Taiwan
Harald Kitzmann	University of Tartu, Estonia
Hassan Silkan	University Chouaib Doukkali, Morocco
Hicham Mouncif	Sultan Moulay Slimane University, Morocco
Insaf Bellamine	Chouaib Doukkali University, Morocco
Ismail Khalil	Johannes Kepler University Linz, Austria
Jemal H. Abawajy	Deakin University, Australia
Jilali Antari	Ibn Zohr University, Morocco
Karim Benouaret	Université Claude Bernard Lyon 1, France
Khalid Housni	Ibn Tofail University, Morocco
Manhua Liu	Shanghai Jiao Tong University, China
Meryeme Hadni	Sidi Mohamed Ben Abdellah University, Morocco
Mhamed Outanoute	Sultan Moulay Slimane University, Morocco
Mohamed Baslam	Sultan Moulay Slimane University, Morocco

Contents

Decision Support and Artificial Intelligence

Optimization Focused on Parallel Fuzzy Deep Belief Neural Network for Opinion Mining

Fatima Es-sabery[1](✉) [ID], Khadija Es-sabery[1] [ID], Bouchra El Akraoui[2] [ID], and Abdellatif Hair[1] [ID]

[1] Department of Computer Science, Faculty of Sciences and Technology, Sultan Moulay Slimane University, 23000 Beni Mellal, Morocco
fatima.essabery@gmail.com

[2] Department of Mathematics, Faculty of Sciences and Technology, Sultan Moulay Slimane University, 23000 Beni Mellal, Morocco

Abstract. In this work, we propose a new parallel fuzzy deep belief neural network for sentiment analysis. We have applied several preprocessing tasks to enhance data quality and remove noisy data. Then, we have applied a semi-automatic data labeling over the dataset by combining two techniques: Vader lexicon and Mamdani's fuzzy system. In addition, we have used four extraction techniques, which are: TFIDF (Unigram), TFIDF (Bigram), TFIDF (Trigram) and GloVe in order to represent each tweet by numerical vector. Further, we have implemented three feature selection techniques which are: The mutual information approach, the chi-square method and the ANOVA technique. Finally, we have applied the deep belief network as classifier in order to classify each tweet into a neutral, negative or positive and our hybrid parallel deep-fuzzy belief neural network is deployed in a parallel design employing the Hadoop framework to overcome the issue of long runtime of huge data sets. Also, a comparisons of the proposed model's effectiveness with other existing models in the literature is carried out and the experimental results shown that our suggested parallel fuzzy model surpasses the baseline models by a considerable margin in terms of recall, runtime, F1 score, accuracy, error rate and precision.

Keywords: Deep belief neural network · Sentiment analysis · Hadoop · HDFS · MapReduce · Selectors of features · Extractors of features · Fuzzy logic

1 Introduction

Sentiment analysis is an automatic language processing and information extraction task. For a given text, the polarity of the text must be identified as either positive, neutral, or negative and give several methods, performance benchmarks and resources to accomplish this task [1]. The polarity of a feeling can be calculated according to several thresholds and can be seen as several different classes [2].

© Springer Nature Switzerland AG 2022
M. Fakir et al. (Eds.): CBI 2022, LNBIP 449, pp. 3–28, 2022.
https://doi.org/10.1007/978-3-031-06458-6_1

Sentiment analysis is an extremely active research area in automatic language processing. Indeed, the last few years have seen an increase in the number of sources of opinion-based textual data available on the web: Internet users' opinions, which are increasingly centralized by search engines, forums, social networks, consumer surveys conducted by major brands [3]. Faced with this abundance of data and sources, automating the synthesis of multiple opinions becomes crucial to efficiently obtain an overview of opinions on a given subject. The interest of this data is considerable, for companies wishing to obtain customer feedback on their products or their brand image as well as for people wishing to inquire about a purchase, an outing, or a trip.

Recently, micro-blogging platforms are attracting the attention of users and researchers because of the ease and speed of information sharing [4]. These platforms can be considered as a very large repository of information containing millions of text messages generally organized in complex networks involving users interacting with each other at specific times. Due to its huge popularity, Twitter is considered the number one micro-blogging platform in the world, it offers APIs that allow to collect free data that can be used to develop applications or perform analysis, that's why we chose it for our different experiments [5].

Numerous studies have been done on Twitter sentiments data in various fields: marketing, politics, natural disasters, etc. Indeed, today, Twitter is one of the best opportunities available for a brand to gain visibility among potential consumers [6]. Marketers are taking note of the many opportunities offered by Twitter and are starting to implement new social initiatives at an unprecedented rate. As a result, global companies have recognized Twitter as a marketing platform in its entirety, and use it with innovations to fuel their advertising campaign [7]. Twitter is also being exploited as a platform for political campaigns by becoming an integrated media at the heart of the political communication strategy [8].

From the point of view of opinion mining methods, this freedom of expression is a major difficulty, because the objective is to find the respondents' concerns within unstructured texts. This justifies the important work done on this subject in the field of automatic natural language processing, suitable for solving this type of data extraction [9]. Indeed, the noisy nature of these texts, characterized by the presence of spelling mistakes, intentional or not, syntactic errors or content formatting, is a challenge at several stages of text analysis, from preprocessing (sentence and word segmentation, grammatical categorization and lemmatization) to term and opinion extraction [10].

Deep learning models can be employed to discover precious information that are hiding in the raucous social media content produced on a daily basis [11]. There are several deep learning models that assist in learning, like Long Short Term Memory Networks (LSTMs), Radial Basis Function Networks (RBFNs), Convolutional Neural Networks (CNNs), Generative Adversarial Networks (GANs), Recurrent Neural Networks (RNNs), Deep Belief Networks (DBNs), Self Organizing Maps (SOMs), Multilayer Perceptrons (MLPs), Autoencoders and Restricted Boltzmann Machines (RBMs). These deep learning approaches operate on almost any kind of data and need high levels of computational performance and knowledge to tackle complicated issues [12].

In this contribution, we developed a new parallel hybrid fuzzy deep learning model, which integrates the advantages of Mamdani's fuzzy system and deep belief networks to perform sentiment classification with high performance. Furthermore, this contribution is implemented in a parallel manner using the Hadoop framework with its HDFS distributed file system and the MapReduce programming model. Therefore, the main propositions of our contribution can be summarized as follows:

1. Our parallel hybrid fuzzy deep belief neural network categorizes the gathered tweets into three categories: negative, positive or neutral.
2. Data pre-treatment tasks, like lemmatization process, warping operation, stop words, and negation procedure, are employed to boost the quality of the tweets by suppressing the undesired and noisy data.
3. Semi-automatic data labeling by combining two techniques: Vader lexicon and Mamdani's fuzzy system
4. Four extraction techniques are applied, namely TFIDF (Unigram), TFIDF (Bigram), TFIDF (Trigram) and GloVe in order to represent each tweet by numerical vector.
5. Three feature selection techniques are implemented, namely mutual information approach, chi-square method and ANOVA technique.
6. Deep Belief Network is used as classifier in order to classify each tweet into a neutral, negative or positive.
7. Our hybrid parallel deep-fuzzy belief neural network is deployed in a parallel design employing the Hadoop framework to overcome the issue of long runtime of huge data sets.
8. Comparisons of the proposed model's effectiveness with other existing models in the literature.
9. Our suggested parallel fuzzy model surpasses the baseline models by a considerable margin in terms of recall, runtime, F1 score, accuracy, error rate and precision.

The remaining part of this proposal is being formed as follows: the second section outlines the previous research works, the third section presents the materials and methods of our proposal, the fourth section discusses the obtained findings, and finally, the "Conclusions" section synthesizes the suggested methodology and makes some recommendations for future work.

2   Previous Research

The following are examples of research papers that employed several different deep learning models to address sentiment analysis issues in a variety of languages. Chen et al. [13] proposed a divide-and-conquer model that first sorts phrases into distinct kinds, and then carries out separate opinion mining on phrases of each kind. In particular, the authors observed that phrases are inclined to be more complicated if they include more emotional labels. Hence, they proposed to firstly implement a neural network-based sequential pattern to categorize the opinionated phrases into three kinds according to the number of targets which appear in a phrase. Each cluster of phrases is then fed independently into a one-dimensional CNN for opinion classification. Their hybrid model was tested on four different opinion classification datasets and was compared to a large set of baselines approaches.

The authors of the paper [14] have integrated the advantages of CNNs and LSTMs to perform the Arabic opinion mining on a diverse set of data. They have used CNN to detect and pick the appropriate features and LSTM to learn the sequential features. They also investigated the efficiency of employing several levels of opinion mining due to the morphological and orthographical complexity of Arabic and they have applied word-level and Ch5gram-level methods to raise the set of extracted features for each tweet and both approaches demonstrated significantly improved opinion classification performance. Their hybrid model enhanced the precision of opinion classification for the Arabic Health Services (AHS) dataset to attain 0.9568% for the Sub-AHS dataset and 0.9424% for the Main-AHS dataset.

In the work [15], a bidirectional deep CNN-RNN design with attention as the basis is proposed. This approach overcomes two drawbacks of DNNs, namely the high dimension of the feature space and the equal importance rate assigned to all features by integrating two separate bidirectional LSTM and GRU layers. Which offers the possibility to extract both past and future features by taking into account the time flow of the data in both directions. In addition, the attention mechanism is implemented on the outcomes of the two-way layers to provide more or less importance on different terms. In order to minimize the feature dimensionality and capture local position-invariant features, the hybrid model employs both convolution and pooling techniques. The results of comparing their hybrid model with six more recently suggested DNNs for opinion mining demonstrate that their hybrid model gets state-of-the-art performance in classifying the polarity of long comments and short tweets.

In [16], the authors propose an innovative approach to learning about real-world relationships with opinion mining of Twitter data using deep learning approaches. With the suggested approach, it is feasible to forecast the satisfaction of the consumer of a product, his enjoyment in a specific environment or the situation of destruction after a disaster. The main reason to use CNN for sentiment classification and analysis is that CNN can detect and capture a set of feature from comments or tweets and is capable of taking into account the relationship between those features. The experimental findings demonstrate that their approach obtains better classification rate in classifying Twitter sentiments than some conventional approaches such as NB and SVM techniques.

Vo et al. [17] integrate Convolutional neural network (CNN) and Long Short Term Memory (LSTM) to perform Vietnamese sentiment analysis corpus. In their work, CNN and LSTM are used to produce channels of data for Vietnamese opinion analysis. Since every deep neural network model (e.g., CNN, LSTM) has a specific benefit, this scenario offers a new and efficient manner to incorporate the advantages of CNN and LSTM. Furthermore, they provided a Vietnamese corpus, which gathered reviews/comments from Vietnamese commercial websites and was annotated in three class label positive, neutral, or negative. From the experimental findings, the suggested model surpasses CNN, SVM, and LSTM on Vietnamese and VLSP datasets.

3   Materials and Methods

In the following subsection of this paper, we will introduce the causes that motivate us to suggest and develop this hybrid model. In general, the basic structure of this suggested paradigm is made up of six stages; The former task is data gathering in which we employed the Sentiment140 huge dataset to assess our suggested hybrid model. The second phase, known as data pretreatment, aims to suppress undesired and noisy data. The third step is the semi-automatic data labeling over the dataset by combining two techniques: Vader lexicon and Mamdani's fuzzy system. The fourth step is data extraction, which turns the tweets into numerical vectors. The fifth stage is feature selection to decrease the dimensions of the retrieved characteristics. Finally, we have implemented Deep Belief Network as classifier for every tweet to be classified in three class labels (neutral, negative or positive). In addition, our hybrid parallel deep-fuzzy belief neural network is deployed in a parallel design employing the Hadoop framework to overcome the issue of long run-time of huge data sets.

3.1   Motivation

Currently, e-commerce platforms permit their customers to post critiques or feedback on the items they have bought. Information provided by customer feedback is crucial in assisting other prospective customers to make a decision on whether or not to purchase a product on the basis of the experiences and opinions of other customers on a specific item. Opinions about a particular product. In addition, manufacturers can also collect user feedback from online reviews to improve their products. However, with the rise in the number of customers buying products, the number of reviews also increases over time and it is not possible for users or manufacturers to read all the reviews to get feedback from previous customers on a certain product. In addition, some of the reviews are lengthy, making it difficult for users to recognize good and bad product features when deciding if the product is worth buying; or for manufacturers to decide if the product needs improvement. An opinion analysis process, which can analyze whether a customer gives a positive or negative review of a certain item, is important and highly recommended for prospective customers and producers because it allows them to easily gather valuable product details from a variety of comments, which helps them in making decisions on the basis of the opinions of others. Motivated by the considerable influence of opinion mining in our daily life. In this contribution, we have proposed a new parallel hybrid deep and fuzzy belief neural network that is used to perform sentiment classification. This hybrid model integrates NLP tasks for carrying out the data pretreatment,

Vader lexicon and Mamdani's fuzzy system for semi-automatic data labeling over the dataset, representation methods to transform every tweet into numerical vector, feature selection approaches to get the most pertinent characteristics and reduce the high dimensional feature vector space, Deep Belief Network as classifier for categorizing every tweet into three targets (neutral, positive or negative). Finally, the Hadoop framework to overcome the issue of long run-time of huge data sets.

The former phase of our proposal is the data pretreatment tasks. Thus, pretreatment has an important impact on the process of text classification, as described in this paper [18]. Their authors have given us a comparison to evaluate the effect of data pretreatment techniques on the classification of tweets in terms of classification rate. The empirical finding has shown that the application of the text pretreatment tasks on lingual text achieved a significant enhancement in the classification performance. Numerous literature works [19,20] demonstrated that the pre-treatment techniques are positively impacted the classification process in terms of classification rate. We were therefore encouraged by the good results reported about the data pretreatment techniques, and we have implemented these pretreatment tasks in this contribution.

After the pretreatment stage, the next stage is the semi-automatic data labeling over the dataset by combining two techniques: Vader lexicon and Mamdani's fuzzy system. The next stage is the data representation in which we applied four extraction techniques namely TFIDF (Unigram), TFIDF (Bigram), TFIDF (Trigram) and GloVe in order to represent each tweet by numerical vector. Then, we implemented three feature selection techniques namely mutual information approach, chi-square method and ANOVA technique. In the final stage, the deep belief network is used as classifier in order for classifying every tweet into a neutral, positive or negative class label and our hybrid parallel deep-fuzzy belief neural network is deployed in a parallel design utilizing the Hadoop framework.

In conclusion, the objective of this study is to increase the classification effectiveness of sentiment analysis by incorporating the strong points of data pretreatment tasks in enhancing the tweets correctness by eliminating the undesired and noisy data, of the feature extractors that converts every tweet into digital vector and retrieves the most pertinent features, of the feature selectors for decreasing the highly dimensional features retrieved in the preceding stage and choosing the most attractive features, Also it integrates the strong points of deep belief neural network in classifying the entry tweet and enhancing the classification effectiveness and that of the Hadoop framework to overcome the issue of long runtime of huge data sets.

As depicted in Fig. 1, our hybrid parallel deep-fuzzy belief neural network's global system is composed of five phases: tweets gathering step, tweets pretreatment step, tweets representation step, feature extraction step, feature selection step, tweets classification, tweets parallelization using Hadoop.

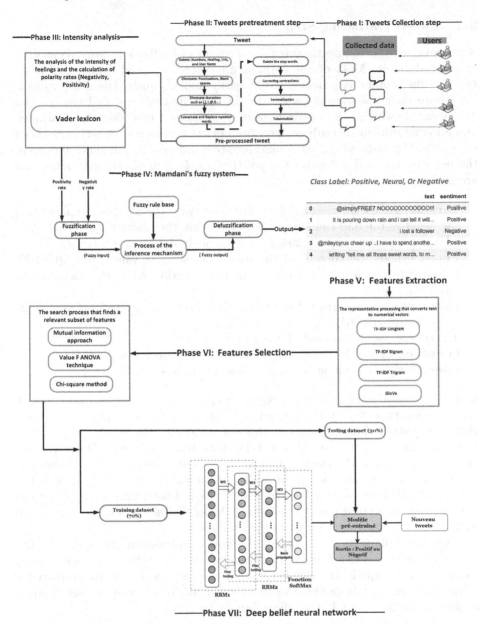

Fig. 1. Global architecture of the proposed hybrid parallel deep-fuzzy belief neural network.

3.2   Tweets Gathering Step

Sentiment analysis applications require a corpus of feedback to learn a classifier or to evaluate it. Most of the corpora used were collected from social sites, because the content is provided freely, easily and instantly. Users can express and share their opinions in public. In this contribution, we used the corpora Sentiment140 which is a massive dataset. It contains 1 600 000 tweets gathered and all emoticons in this sub-set have been deleted. For every tweet, it has been categorized by employing 2 class tags: negative and positive, in which 0 denotes the negative tag and 4 denotes the positive tag. It includes six attributes that are listed as follows:

- **Target:** Identify the class tag for every tweet, with the number 0 representing the negative tag and the number 4 representing the positive tag.
- **Ids:** is a number unique (520442) that uniquely defines every tweet.
- **Flags:** defines the content of the user' request. The value "NO QUERY" is attributed to flags characteristic in the situation when the user doesn't publish a request.
- **Date:** shows the correct time when the tweet was posted (Mon Jan 26 04:34:58 +0101 2001).
- **Text:** presents the content of every tweet.
- **Location:** shows the correct location where the tweet was posted.
- **User:** denotes the name of the user who has published the tweet.

In this work, we concentrate on exploring opinions. This implies that we get each opinion provided by the individual Twitter user in each published tweet. Hence, the other characteristics of this dataset do not have any effect on the training task. Therefore, we discarded the characteristics "User", "Date", "Ids", "Flag" and retained the characteristics "Text" and "Target" from the dataset. In this contribution, the chosen dataset is partitioned into 2 sub-sets, namely the formation and test sub-sets. Therefore, we employed both subsets to exhibit the efficiency of our hybrid parallel deep-fuzzy belief neural network in comparison with other published approaches in the literature.

Figure 2 displays the count of neural, positive and negative tweets within the formation and test sub-set. In the formation procedure, we have employed a total count of tweets equals 1120000 tweets. For the test procedure, we employed a count of tweets equals 480000 tweets. In other words, the test sub-set is equal to 30% of all the tweets in the whole dataset.

3.3   Tweets Pretreatment Step

The preprocessing task is widely studied and used in many applications that take into account raw and unstructured data. When research focused more on comments in social networks, the need for pretreatment is raised proportionally, because numerous comments contain spelling errors and are not properly formatted [21]. Therefore, preprocessing techniques are needed to acquire a cleaner

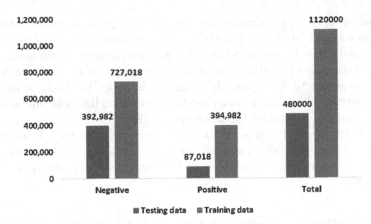

Fig. 2. Count of neutral, negative, and positive tweets for the Sentiment140 dataset.

dataset. Dataset cleaning significantly improves the performance rate of the subsequent classification task.

In almost all of the opinion mining work that has been mentioned [9–20], the preprocessing techniques used are incorporated The techniques applied are straightforward, such as filtering of words, letters, punctuation and correction of simple errors. The correction of simple errors, such as repeated letters and spelling mistakes is based on lexicons. Dictionaries are utilized to correct errors. Similarly, acronyms and abbreviations are replaced by terms from a well-defined lexicon.

An alternative approach that has been employed is eliminating unnecessary content. Such as punctuation and stop words do not have a strong impact on the sentimental score, they are withdrawn to diminish the variety of words utilized in the tweets. An instance of a filtering strategy is the excessive usage of letters. An example of vowel repetition is "Weeeeeeeell". An example of punctuation repetition is "well !!!!!!!!!!!!!!!!". These filtering techniques can be performed by identifying the excessive usage of more than 2 of the following letters [22].

The regular expression is the most popular approach for replacing and filtering strings. It allows to identify and to remove the abuses and the errors, and more commonly, it offers a useful tool for recognizing sub-strings in a string. Regular expressions have been deemed to be a very strong matching mechanism for strings, which is applied in numerous "search and replace" functionalities of applications [23].

There is another valuable strategy, termed Tokenization, that can be employed to minimize the large diversity of terms. Tokenization is a procedure for reducing verb forms to the basic stem of the verb. An instance of Tokenization is the simplification of "liked" to "like". This minimizes the diversity of verb conjugations in which a term can appear and therefore, minimizes the quantity of the data [24].

Most of the techniques outlined are specific to a given language. The text's language has an essential role to play in preprocessing and classification. It is crucial to obtain the cleanest possible dataset. If we examine multiple languages, every one of them has a distinct grammatical form, structure, and term lexicon. It is thus necessary to determine the language in order to choose a particular linguistic context. This is the primary field of research on linguistic identification, which has been the subject of extensive studies [25].

Once the language of a message has been identified, the model that is to be trained can assume that all messages come from the identified language. It can therefore be assumed that only words of the identified language appear in the messages [26].

Data Exploration: A classical data representation in automatic language processing is the word cloud. In this form, the most commonly used words can be visualized very easily [27]. Apart from the colors, which have a purely decorative function, the frequency of the words in the database used is proportional to the size of the words in question in the image (as shown in the Fig. 3).

Before pretreatment   After pretreatment

Fig. 3. Word cloud.

Here, we have plotted the raw data with and without any preprocessing on the two provided graphical representations. In the case of raw data, the observed noise is mainly due to the structure of the tweet headers. In the case of preprocessed data (slightly), we manage to filter out a good part of the noise and the data becomes informative and can start to be analyzed.

In this contribution, the stage that follows the tweets pretreatment step is the semi-automatic labeling step. This means that the text returned after employing all the tweets pretreatment tasks previously outlined will be feed to the combined technique (Mamdani fuzzy system+Vader lexicon).

3.4   Semi-automatic Labeling Step

Data Labeling is an essential step in Machine Learning. To train an AI model from data, it is imperative to first label the data. After a thorough study and analysis of the dataset, we observed that there are some tweets that are mislabeled therefore we decided to do a semi-automatic data labeling by combining two techniques: Vader lexicon and Mamdani fuzzy system

Vader Lexicon. Valence Aware Dictionary and sentiment Reasoner (VADER) is rule-based linguistic database and opinion mining process that is designed to analyze social network emotions. It is an open source under the MIT license developed by George Berry, Ewan Klein, and Pier Paolo. It employs a diversity of tools and technologies. A sentiment dictionary is a set of lexical characteristics (e.g., words) that are categorized as negative or positive based on their polarity. It presents not only the negativity and positivity rates, but also the intensity of a negative or positive feeling. VADER preserves the advantages of conventional lexicons like LIWC (Linguistic Inquiry and Word Count). It is larger, easy to control, simple to comprehend, fast to apply and easy to expand (Table 1).

Table 1. Polarity intensity.

Tweet	Positivity rate	Negativity rate
The deep neural networks are very efficiency to process data; but they are inefficiency with the ingrained ambiguity in NL that needs more solutions	0.575	0.425

Mamdani's Fuzzy System: After calculating the two values TP (Positivity rate) = 0.575 and TN (Negativity rate) = 0.425. The next phase is the application of the Mamdani fuzzy system which is based on three steps as shown in the Fig. 4.

Fuzzification Phase: The first step is the application of fuzzification process to the net values of TP and TN, using the triangular membership function (described by the Eq. (4)) to compute the degree of membership μ of the TP and TN to the fuzzy sets Weak, Middle, and High.

$$\mu_A(x) = \begin{cases} 0 & \text{si } x \leq ll \\ \frac{x-ll}{v-ll} & \text{si } ll \leq x \leq v \\ \frac{ul-x}{ul-v} & \text{si } v \leq x \leq ul \\ 0 & \text{si } c \leq x \end{cases} \tag{1}$$

where ll is the lower bound, ul is the upper bound, the median value v, and $ll < v < ul$.

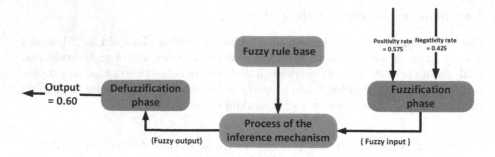

Fig. 4. Phases of Mamdani fuzzy system.

Fuzzy Rule Base: After the fuzzification process, the following stage is the setting up of the fuzzy rules IF-THEN.

Rule 1: IF TP is Weak **AND** TN is Weak **THEN** CL = neutral
Rule 2: IF TP is Weak **AND** TN is Middle **THEN** CL = Negative
Rule 3: IF TP is Weak **AND** TN is High **THEN** CL = Negative
Rule 4: IF TP is Middle **AND** TN is Weak **THEN** CL = Positive
Rule 5: IF TP is Middle **AND** TN is Middle **THEN** CL = Neutral
Rule 6: IF TP is Middle **AND** TN is High **THEN** CL = Negative
Rule 7: IF TP is High **AND** TN is Weak **THEN** CL = Positive
Rule 8: IF TP is High **AND** TN is Middle **THEN** CL = Positive
Rule 9: IF TP is High **AND** TN is High **THEN** CL = Neutral

Inference Mechanism: After having defined the fuzzy IF-THEN rules. The following phase is the application of the inference process that consists of three rules: Application, Implication, and Aggregation.

Defuzzification Phase: The ultimate stage of Mamdani's fuzzy system is the Defuzzification by applying the centroid method described by the following equation:

$$d_v = \frac{\sum_{i=1}^{n} z_i . \mu(z_i)}{\sum_{i=1}^{n} \mu(z_i)} \tag{2}$$

where z_i denotes the instance element, $\mu(z_i)$ is the degree of membership of the element z_i, and n describes the number of elements in the instance.

3.5   Tweet Representation Step

For many machine learning models, it is necessary to represent the features as vectors of real numbers, because the feature values must be multiplied by the model weights, which is why feature extraction is used [28]. In our work, we have used 4 extraction techniques, namely TFIDF (Unigram), TFIDF (Bigram), TFIDF (Trigram) and also GloVe in order to find out which one offers high performance accuracy.

N-grams: is a sub-sequence of n elements built according to a defined sequence. The concept seems to come from Claude Shannon's contribution in theory of information. His concept was that, based on a chosen sequence of characters for example, it is then easy to get the probability function of the apparition of the following character. Based on a sample corpora, it is straightforward to compute a probability distribution for the subsequent character using a background of dimension n. N-grams method is extensively applied in NLP. Its usage relies on the simplification hypothesis that, given a string of k characters ($k \geq n$), so the likelihood of the occurrence of a characters (OC) at position i is related only to the n-1 preceding characters.

TF-IDF: TF-IDF is a strict statistical method that relies on the occurrence of terms. It is commonly employed particular in data mining and in data retrieval. This statistical metric is designed to assess the relevance of a term within a document, relative to a corpus or a collection. The weight rises proportionally the frequency with which the term appears in the document. It also changes according to the number of times the term is present in the corpora. Variations of the original formula are often used in research engines to assess the pertinence of a document according to the user's search criteria.

GloVe: When employing word integration patterns, words are mapped into a vector space with real values. A proper integration of words should preferably map words in such a manner that two distinct words with very nearly the same semantic significance have very similar mappings in the vector space. Other lingual connections between dissimilar words can also be retained. For example, when we use these representations of the vector space for the words, the following operations "King - Man + Woman" give a score very close to the representation of the vector space for the word "Queen". We try to represent each word i and each word j matched in the same context by vectors v_i and v_j respectively, of dimension d such as:

$$v_i.v_j + b_i + b_j = log(X_{ij}) \qquad (3)$$

where X_{ij} represents the number of times that the word j occurs in the context of word i. b_i and b_j are scalar biases associated with words i and j respectively.

3.6 Feature Selection Step

Characteristic or feature selection is a research procedure aimed at finding the most appropriate subset of features from the starting set [29]. The notion of relevance of a subset of features always depends on the objectives and criteria of the system. In our work we have performed the feature selection by combining three methods:

Mutual Information Approach: In information and probability theories, the mutual information of 2 random variables is a measurement of the relative relationship between the 2 random variables. More precisely, it is the amount of information that is gathered about one random variable by examining the other random variable. It is symmetric and can detect non-linear relationships between both random variables.

Chi-square Method: The chi-square approach is mainly utilized in statistics to test the independence of two events. It is a digital test which calculates the deviation with respect to the predicted distribution by taking into account that the feature event is not correlated with the value of the decision feature. Chi-square in feature selection calculates whether the presence of a particular term and the presence of a particular class are both independent. Thus each term is evaluated and all terms have been ordered by their score. A high score indicates that the null assumption of importance must be discarded and that the presence of the term and the class are linked to each other. If the class and the term are dependent on each other, the feature is chosen for classification.

ANOVA Technique: is an acronym for "analysis of variance". This is a parametric statistical of the hypothesis test to determine whether or not the means of two or more samples of data (often three or more) are from the same distribution.

3.7 Tweet Classification Step

Deep Belief Network (DBN) is a concatenation of several restricted Boltzmann machines (RBM). An RBM is trained in an unsupervised manner and used as a feature extractor [30]. This model has the ability to handle complicated data structures and to detect features that cannot be detected directly. Therefore, stacking RBMs in a successive manner can represent more complex structures. Indeed, the output of the previous RBM is used to train the current RBM to detect features implicit in the previous RBM, etc. In general, the DBN is trained using the layer-by-layer algorithm, its advantage is to discover descriptive features that illustrate the correlation between the inputs in each layer. The layer-by-layer learning algorithm allows for better optimization of weights between layers. In addition, initializing the DBN weights could improve the results rather than if random weights are used. Furthermore, the benefits of DBN learning are extended to its ability to decrease the effects of overlearning and underlearning, where both are common problems in large deep learning architectures. For these reasons, and in this paper, the DBN is chosen as the classifier.

Cross-Validation Technique. Cross-Validation (CV) is the most commonly used approach for setting hyper-parameters, that is discussed in the survey [33]. In the case of the 10-fold CV, like the popular CV but for every hyper-parameter adjustment, ten performance measure rates are computed. Afterwards, the average achievement metric is computed for every hyper-parameter. The greatest average performance metric is used as the ultimate outcome measure for the system. In our contribution, the hyper-parameters of the DBN model were adjusted based on the values donated in the Table 2:

Table 2. Setting the parameters of our DBN model.

Parameters	Values
Learning rate	0.0001
Activation function on the output layer	Sigmoid
Activation functions in the hidden layers	Relu
Number of layers	4–6
Batch size	64
Number of epochs	25
Output	1
Learning rate of RBM	0.0015
Hidden neurons per hidden layer	185–200
Number of iterations back-propagation	100

The choice of hyper-parameters, like the depth of features, the batch size, the number of hidden units, the number of visible and hidden layers, the both learning rate DBN and RBM, and the number of epochs, has a great impact on the accuracy of the classification and the complexity of the calculation. If the depths are incorrectly defined, the accuracy of DBN may not be better than conventional machine learning methods. To find the optimal number of hidden and visible RBM layers and the optimal number of hidden cells to employ, all combinations of values must be explored. In this contribution, we have employed the grid search approach for the parameter selection. The grid search approach is mainly an efficient method to choose the best values for the hyper-parameters of a particular algorithm that attain the better performance. It is employed in this contribution with a CV of 10 fold to pick the optimal hyper-parameters for the DBN pattern. Nevertheless, our experimental findings give some indication of trustworthy ranges for our DBN hyper-parameters, i.e., approximately 4 to 6 RBM hidden layers with 185 to 200 hidden neurons per hidden layer seems to be sufficient, the batch size is 64, the DBN learning rate equals 0.0001, the RBM learning rate equals to 0.0015, the number of epochs is 25 and the number of iterations back-propagation equals 100.

3.8 Parallelization Using MapReduce Version 2 with Hadoop

In this work, we have employed the Hadoop framework in our proposed approach to reduce the runtime and enhance the performance of our proposal. Our suggested methodology implicates the utilization of a huge dataset (Sentiment140). In the first stage, we have utilized HDFS to stock and distribute the huge dataset in the parallel manner among all the computational computer of our Hadoop cluster. After storing the dataset in HDFS, the following stage is the application of our proposed approach over the stored dataset. In this second stage, we have implemented the MapReduce scheduling pattern for parallelizing our approach across all computational computer in the Hadoop cluster [31,32]. The entry to each cycle of the

MapReduce algorithm is a tweet to be classed, and the result is a classed tweet with the decision of classification. The result of the classification of every tweet will also be saved in the HDFS as described in the Algorithm 1 which introduces the MapReduce algorithm implemented in our proposal to classify the tweets.

3.9 Parallelization Using Apache Spark with Hadoop

Apache Spark is a novel infrastructure that can offer an enhanced version of the MapReduce pattern. In contrast to the MapReduce pattern, Spark does not clear the data to the drive at every stage, but instead the data is treated in storage memory till the storage of the memory is reached. When the memory is used up, then it dumps the data to the hard drive. Consequently, Spark can also be termed in-memory computing. This benefit of Spark can enable it to be handled very speedily when compared to MapReduce pattern. Spark is commonly reported to be up to 10 times quicker than the MapReduce pattern. The Spark platform can be deployed in different ways such as HBase, Yarn, Standalone, or Cassandra. This is another benefit of spark that, in contrast to MapReduce, does not used HDFS to operate, but instead spark can operate on a variety of other ways as noted above. The structure of Spark is illustrated in Fig. 5. The Spark platform utilizes a master/worker configuration in which the worker nodes are controlled by the master node. In this proposal, we have used both MapReduce and Spark under Hadoop in order to compare them in term of the runtime of our proposed approach.

3.10 Training and Evaluation Data Set

After the feature selection phase, we divided the data set into two subsets:
Learning set: subset for learning and training the model.
Evaluation set: subset used to evaluate the model.
To train our model we have taken 70% of the dataset and 30% to evaluate it (Table 3).

3.11 Evaluation Criterias

Evaluating the performance of a classification system is a very important task. There are several indicators to measure the performance of models, Each one has its own specific features and it is often necessary to use several of them to have a complete vision of the performance of our model. In this work, we have used four evaluation criterias which are [31]:

Precision: measures the accuracy of a classifier. Higher accuracy means fewer false positives, while lower accuracy means more false positives. This measure is calculated using the following Eq. (4).

$$Precision = \frac{vp}{vp + fp} \tag{4}$$

Algorithm 1: Our MapReduce fuzzy deep belief algorithm

Input   : Dataset of tweets
Output: decision of the classification.
Set up the task of Hadoop
Define the SelectJobMapper as a Mapper category
Define the selectJobReducer as the Reducer category
Adapt the size of the HDFS block till the dataset **D** can be broken down into **P**
  partitions $D_k = \{k = 1, 2, ..., P\}$
In the k-th SelectJobMapper
Input   : $D_k = \{d_1, d_2, ..., d_p\}$ is the non labelled dataset
Output: (key,value)=(d_k, ClassLabel)
for $k \leftarrow 1$ **to** P *partitions* **do**
 | Tweets pretreatment step
 | Using the Vader lexicon for computing:
 | 1. Positivity rate (TP)
 | 2. Negativity rate (TN)
 |
 | Using the Mamdani fuzzy system by applying the following steps:
 |
 | 1. Fuzzification stage
 | 2. Application sub-stage
 | 3. Implication sub-stage
 | 4. Aggregation sub-stage
 | 5. Defuzzification
 |
 | **MapperOutput** : (key,value)=(d_k,ClassLabel)
end for
In the k-th SelectJobReducer
Input   : (key,value)=(d_k, $list(ClassLabel_i)$)
Output: (key,value)=(d_{k*}, $ClassLabel_i^*$
for $k \leftarrow 1$ **to** P *Partitions* **do**
 | ClassLabel$_i^*$ = $\sum_{j=1}^{p} ClassLabel_j$.
end for
In the k-th SelectJobMapper
Input   : $D_k^* = \{d_1^*, d_2^*, ..., d_p^*\}$ is the labelled dataset
Output: (key,value)=(d_k^{**}, $ClassLabel^{**}$)
for $k \leftarrow 1$ **to** P *partitions* **do**
 |
 | 1. Tweet representation step
 | 2. Feature selection step
 | 3. Tweet classification step
 |
 | . **MapperOutput** : (key,value)=(d_k^{**}, $ClassLabel^{**}$)
end for
In the k-th SelectJobReducer
Input   : (key,value)=(d_k^{**}, $list(ClassLabel_i^{**})$)
Output: (key,value)=(d_k***, $ClassLabel_i***$
for $k \leftarrow 1$ **to** P *Partitions* **do**
 | ClassLabel$_i^{***}$ = $\sum_{j=1}^{p} ClassLabel_j^{**}$
end for
return *Classification decision: ClassLabel*$_i^{***}$

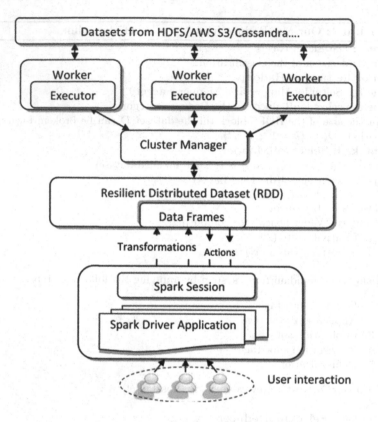

Fig. 5. Global structure of Spark.

Table 3. Training and evaluation data set.

Learning set	Evaluation set
1120000	480000

Recall: measures the completeness, or sensitivity of a classifier. Higher recall means fewer false negatives, while lower recall means more false negatives. This measure is calculated using the following Eq. (5).

$$Recall = \frac{vp}{vp + fn} \tag{5}$$

F1-Score: is the weighted harmonic mean of the accuracy and recall. This measure is calculated using the following Eq. (6).

$$F1 - Score = 2.\frac{Recall.Precision}{Recall + Precision} \tag{6}$$

Classification rate: allows to measure the performance of the model on positive and negative items in a symmetrical way. It measures the rate of correct predictions on all items. This measure is calculated using the following Eq. (7):

$$ClassificationRate = \frac{tp + tn}{tp + fp + tn + fn} \tag{7}$$

where:
tp: Number of tweets that are actually positive and expected to be positive.
tn: Number of tweets that are actually negative and expected to be negative.
fp: Number of tweets that are actually negative and expected to be positive.
fn: Number of tweets that are actually positive and expected to be negative.

4 Results and Discussion

In this section we will discuss the results obtained after several experiments. The first experiment is used to evaluate the influence of each step of the pre-treatment stage on the dataset, the second one aims at finding the most efficient feature extractor, the third one aims at detecting the higher performance feature selector and the last one shows the results of different combination performed in our approach.

4.1 Dataset After Labeling

As presented earlier, the tweets in the sentiment140 dataset are misclassified because the creators of this dataset assumed that any tweet containing positive emoticons, such as :), is positive, and that tweets containing negative emoticons, such as :(, are negative. Therefore, we decided to re-label by combining the Vader lexicon and Mamdani's fuzzy system which is based on manually created rules. Figure 6 shows the data set after labeling.

4.2 Imbalanced Dataset

A widespread adopted and maybe the simplest approach to handling highly skewed datasets is called resampling. It consists of withdrawing examples of the majority class (under-sampling) and/or adding more examples of the minority class (over-sampling) as shown in the Fig. 7. Figure 8 shows our dataset after the application of under-sampling and over-sampling.

4.3 Evaluation of Feature Extractors

The goal of feature extraction is to convert the input tweets into a set of numerical vectors. In this second experiment, we determined the most efficient feature extractor based on the classification rate among the feature extractors used in this work: TF-IDF Unigram, TFIDF Bigram, TF-IDF Trigram and GloVe.

Table 4 depicts the classification rate obtained by applying TF-IDF Unigram, TF-IDF Bigram, TF-IDF Trigram and GloVe.

As shown in Table 4, TF-IDF Unigram outperforms the other feature extractors in terms of classification rate and execution time since it achieved a classification rate equal to 83.73% and an execution time of 0.32 s.

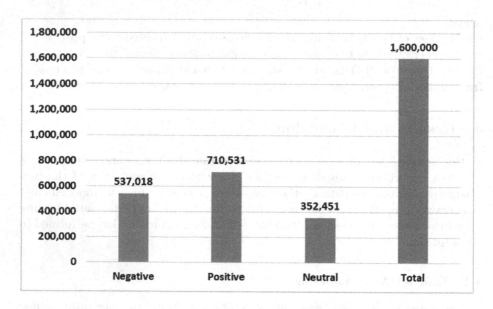

Fig. 6. Data set after labeling.

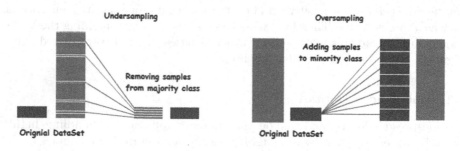

Fig. 7. Under-sampling and over-sampling of the imbalanced dataset.

Table 4. Classification rate and execution time of the used feature extractors.

Feature extractor	Classification rate	Execution time
Tf-Idf Unigram	83.73%	32 s
Tf-Idf Bigram	80.12%	95 s
Tf-Idf Trigram	78%	83 s
GloVe	79%	115 s

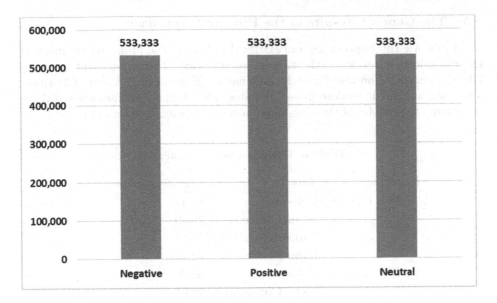

Fig. 8. Under-sampling and over-sampling of the imbalanced dataset.

4.4 Analysis of Feature Selectors

As presented earlier, the step that follows the feature extraction step is the feature selection step. In this phase, we apply many techniques, namely Chi-square, mutual information and analysis of variance, and then we do the hybridization of these three techniques. Therefore, this experiment aims to find the best feature selector among all used selection methods based on classification rate and execution time. Table 5 describes the classification rate and execution time obtained by applying different feature selection techniques.

Table 5. Classification rate and execution time (ET) of the used feature selectors.

Feature selector	Classification rate	ET without Hadoop	ET with Hadoop
Chi-square	67.14%	43.049 s	8.60 s
Mutual information	75.70%	168 s	31.69 s
Variance	57.61%	29.87 s	5.07 s
hybridization	83.73%	240.22 s	34.31 s

As illustrated in Table 5, we notice that the proposed hybridization as a feature selector outperforms other selectors in terms of classification rate since it achieved a classification rate equal to 83.73%.

4.5   The General Results of the Proposed Approach

To emphasize the importance and clarify the effect of the proposed approach in the classification results of the tweets; we performed this experiment to calculate the classification rate for each combination (Extractor+Selector+Classifier) and then analyze it to determine the model with highest performance. Table 6 represents the results of the classification rates for each combination.

Table 6. Classification rate results.

Extractor	Selector	Classification rate (%)
TF-IDF Unigram	Chi-square	67.14
	Mutual information	75.70
	Variance (ANOVA)	57.61
	Hybridization	83.73
TF-IDF Bigram	Chi-square	58.89
	Mutual information	72.01
	Variance (ANOVA)	50.69
	Hybridization	80.12
TF-IDF Trigram	Chi-square	61.05
	Mutual information	69.75
	Variance (ANOVA)	52.14
	Hybridization	78
GloVe	Chi-square	72.17
	Mutual information	50.42
	Variance (ANOVA)	66.10
	Hybridization	79

From the results shown in Table 6, we notice that the combination (TF-IDF Unigram+Hybridization) achieved the highest classification rate (83.73%) as depicted in the Tables 7 and 8.

Table 7. The most efficient classifier.

Classifier	Extractor	Selector	Classification rate
DBN	TF-IDF Unigram	Hybridization	83.73%

Table 8. Examples of the classification performed by the most efficient model on the test set.

Tweet	Actual class	Predicted class
"@simplyFREE7 LOOOOOOOOOOOOO!!!	Positive	Positive
"@mileycyrus cheer up ... I have to spend another ..."	Positive	Positive
"@Zella17 I'm soo sad !!! they killed off kutner on house whyyyyy"	Negative	Negative
"@lucktotheduck Deciding on which new theme to get"	Positive	Positive

4.6   Comparative Study

In this subsection, we conducted a comparison of the results obtained from the proposed approach and those of the Sentiment140 which is part of an automatic sentiment classification project on Twitter. In this project, they performed a classification of tweets using three different ML algorithms which are NB, SVM and Maximum Entropy. In Fig. 9 we summarize all the results obtained from this project and the results of the proposed approach in terms of the evaluation metrics: Accuracy, Recall, F-score and Classification rate.

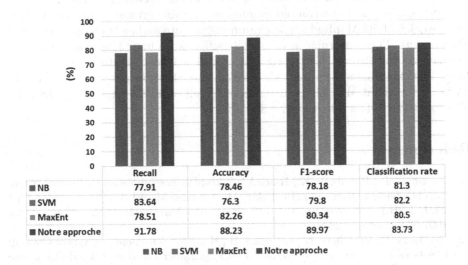

	Recall	Accuracy	F1-score	Classification rate
■ NB	77.91	78.46	78.18	81.3
■ SVM	83.64	76.3	79.8	82.2
■ MaxEnt	78.51	82.26	80.34	80.5
■ Notre approche	91.78	88.23	89.97	83.73

■ NB   ■ SVM   ■ MaxEnt   ■ Notre approche

Fig. 9. Comparative study.

The comparative study allowed us to conclude that the results obtained by our approach outperforms the three algorithms with a high degree, this shows us the advantage of applying a Deep learning approach in this type of sentiment analysis tasks and the good performance we get in case of data massive instead of using traditional machine learning algorithms.

4.7   Runtime of Our Approach

This experiment aims to compute the execution time of our proposed approach when using the MapReduce with Hadoop and when employing the Spark with Hadoop. The obtained result is depicted in the following Table 9. As shown in the Table 9 The execution time when using MapReduce equals 34.31 s and it equals 3.431 s when using Spark. From this result, we deduced that the Spark framework is 10 times faster than the MapReduce pattern.

Table 9. Execution time of our classifier with MapReduce and Spark.

Classifier	Extractor	Selector	ET MapReduce	ET Spark
DBN	TF-IDF Unigram	Hybridization	34.31 s	3.431 s

5   Conclusions

In this contribution, an endeavor has been done to design a new classification pattern to classify the massive dataset samples into their corresponding class labels. Our proposal consists of several stages, which are data pre-processing stage, feature extraction, feature selection, and DBN classifier. The suggested method operates in a distributed manner on scalable groups using the Hadoop framework with its MapReduce programming pattern and its Hadoop distributed file systems. The performance of our classifier is evaluated by computing the evaluations metrics which are precision, recall, F1-Score and the classification rate. The experimental findings shown that our classifier outperforms the NB, SVM and MaxEnt algorithms in terms of Recall = 91.78%, Accuracy = 88.23%, F1-score = 89.97% and Precision = 83.73%.

References

1. Es-sabery, F., Es-sabery, K., Hair, A.: A MapReduce improved ID3 decision tree for classifying twitter data. In: Fakir, M., Baslam, M., El Ayachi, R. (eds.) CBI 2021. LNBIP, vol. 416, pp. 160–182. Springer, Cham (2021). https://doi.org/10.1007/978-3-030-76508-8_13
2. Es-sabery, F., Es-sabery, K., Garmani, H., Hair, A.: Sentiment analysis of Covid19 tweets using a MapReduce fuzzified hybrid classifier based on C4.5 decision tree and convolutional neural network. E3S Web Conf. **297**, 01052 (2021). https://doi.org/10.1051/e3sconf/202129701052
3. Es-Sabery, F., Hair, A., Qadir, J., Sainz-De-Abajo, B., García-Zapirain, B., Torre-Díez, I.D.L.: Sentence-level classification using parallel fuzzy deep learning classifier. IEEE Access **9**, 17943–17985 (2021). https://doi.org/10.1109/ACCESS.2021.3053917
4. Naseem, U., Razzak, I., Musial, K., Imran, M.: Transformer based deep intelligent contextual embedding for twitter sentiment analysis. Futur. Gener. Comput. Syst. **113**, 58–69 (2020). https://doi.org/10.1016/j.future.2020.06.050

5. Carvalho, J., Plastino, A.: On the evaluation and combination of state-of-the-art features in Twitter sentiment analysis. Artif. Intell. Rev. **54**(3), 1887–1936 (2020). https://doi.org/10.1007/s10462-020-09895-6

6. Yi, S., Liu, X.: Machine learning based customer sentiment analysis for recommending shoppers, shops based on customers' review. Complex Intell. Syst. **6**(3), 621–634 (2020). https://doi.org/10.1007/s40747-020-00155-2

7. Botchway, R.K., Jibril, A.B., Oplatková, Z.K., Chovancová, M.: Deductions from a Sub-Saharan African bank's tweets: a sentiment analysis approach. Cogent Econ. Finance **8**, 1776006 (2020). https://doi.org/10.1080/23322039.2020.1776006

8. Hubert, R.B., Estevez, E., Maguitman, A., Janowski, T.: Analyzing and visualizing government-citizen interactions on twitter to support public policy-making. Digit. Gov.: Res. Pract. **1**, 15:1–15:20 (2020). https://doi.org/10.1145/3360001

9. Es-Sabery, F., et al.: A MapReduce opinion mining for COVID-19-related tweets classification using enhanced ID3 decision tree classifier. IEEE Access **9**, 58706–58739 (2021). https://doi.org/10.1109/ACCESS.2021.3073215

10. Sarlan, A., Nadam, C., Basri, S.: Twitter sentiment analysis. In: Proceedings of the 6th International Conference on Information Technology and Multimedia, pp. 212–216 (2014). https://doi.org/10.1109/ICIMU.2014.7066632

11. Zhang, Q., Yang, L.T., Chen, Z., Li, P.: A survey on deep learning for big data. Inf. Fusion **42**, 146–157 (2018). https://doi.org/10.1016/j.inffus.2017.10.006

12. Nosratabadi, S., Mosavi, A., Keivani, R., Ardabili, S., Aram, F.: State of the art survey of deep learning and machine learning models for smart cities and urban sustainability. In: Várkonyi-Kóczy, A.R. (ed.) INTER-ACADEMIA 2019. LNNS, vol. 101, pp. 228–238. Springer, Cham (2020). https://doi.org/10.1007/978-3-030-36841-8_22

13. Chen, T., Xu, R., He, Y., Wang, X.: Improving sentiment analysis via sentence type classification using BiLSTM-CRF and CNN. Expert Syst. Appl. **72**, 221–230 (2017). https://doi.org/10.1016/j.eswa.2016.10.065

14. Alayba, A.M., Palade, V., England, M., Iqbal, R.: A combined CNN and LSTM model for Arabic sentiment analysis. In: Holzinger, A., Kieseberg, P., Tjoa, A.M., Weippl, E. (eds.) CD-MAKE 2018. LNCS, vol. 11015, pp. 179–191. Springer, Cham (2018). https://doi.org/10.1007/978-3-319-99740-7_12

15. Basiri, M.E., Nemati, S., Abdar, M., Cambria, E., Acharya, U.R.: ABCDM: an attention-based bidirectional CNN-RNN deep model for sentiment analysis. Futur. Gener. Comput. Syst. **115**, 279–294 (2021). https://doi.org/10.1016/j.future.2020.08.005

16. Liao, S., Wang, J., Yu, R., Sato, K., Cheng, Z.: CNN for situations understanding based on sentiment analysis of twitter data. Procedia Comput. Sci. **111**, 376–381 (2017). https://doi.org/10.1016/j.procs.2017.06.037

17. Vo, Q.-H., Nguyen, H.-T., Le, B., Nguyen, M.-L.: Multi-channel LSTM-CNN model for Vietnamese sentiment analysis. In: 2017 9th International Conference on Knowledge and Systems Engineering (KSE), pp. 24–29 (2017). https://doi.org/10.1109/KSE.2017.8119429

18. de Oliveira, D.N., Merschmann, L.H.C.: Joint evaluation of preprocessing tasks with classifiers for sentiment analysis in Brazilian Portuguese language. Multimedia Tools Appl. **80**(10), 15391–15412 (2021). https://doi.org/10.1007/s11042-020-10323-8

19. Chintalapudi, N., Battineni, G., Canio, M.D., Sagaro, G.G., Amenta, F.: Text mining with sentiment analysis on seafarers' medical documents. Int. J. Inf. Manag. Data Insights **1**, 100005 (2021). https://doi.org/10.1016/j.jjimei.2020.100005

20. Aljuaid, H., Iftikhar, R., Ahmad, S., Asif, M., Tanvir Afzal, M.: Important citation identification using sentiment analysis of in-text citations. Telemat. Inform. **56**, 101492 (2021). https://doi.org/10.1016/j.tele.2020.101492

21. Soni, V.K., Pawar, S.: Emotion based social media text classification using optimized improved ID3 classifier. In: 2017 International Conference on Energy, Communication, Data Analytics and Soft Computing (ICECDS), Chennai, India, pp. 1500–1505 (2017). https://doi.org/10.1109/ICECDS.2017.8389696

22. Ngoc, P.V., Ngoc, C.V.T., Ngoc, T.V.T., Duy, D.N.: A C4.5 algorithm for English emotional classification. Evol. Syst. **10**(3), 425–451 (2017). https://doi.org/10.1007/s12530-017-9180-1

23. Lakshmi Devi, B., Varaswathi Bai, V., Ramasubbareddy, S., Govinda, K.: Sentiment analysis on movie reviews. In: Venkata Krishna, P., Obaidat, M.S. (eds.) Emerging Research in Data Engineering Systems and Computer Communications. AISC, vol. 1054, pp. 321–328. Springer, Singapore (2020). https://doi.org/10.1007/978-981-15-0135-7_31

24. Guerreiro, J., Rita, P.: How to predict explicit recommendations in online reviews using text mining and sentiment analysis. J. Hosp. Tour. Manag. **43**, 269–272 (2020). https://doi.org/10.1016/j.jhtm.2019.07.001

25. Mehta, R.P., Sanghvi, M.A., Shah, D.K., Singh, A.: Sentiment analysis of tweets using supervised learning algorithms. In: Luhach, A.K., Kosa, J.A., Poonia, R.C., Gao, X.-Z., Singh, D. (eds.) First International Conference on Sustainable Technologies for Computational Intelligence. AISC, vol. 1045, pp. 323–338. Springer, Singapore (2020). https://doi.org/10.1007/978-981-15-0029-9_26

26. Zhang, J.: Sentiment analysis of movie reviews in Chinese. Uppsala University (2020). https://www.diva-portal.org/smash/get/diva2:1438431/FULLTEXT01.pdf

27. López-Chau, A., Valle-Cruz, D., Sandoval-Almazán, R.: Sentiment analysis of twitter data through machine learning techniques. In: Ramachandran, M., Mahmood, Z. (eds.) Software Engineering in the Era of Cloud Computing. CCN, pp. 185–209. Springer, Cham (2020). https://doi.org/10.1007/978-3-030-33624-0_8

28. Addi, H.A., Ezzahir, R., Mahmoudi, A.: Three-level binary tree structure for sentiment classification in Arabic text. In: Proceedings of the 3rd International Conference on Networking, Information Systems & Security (NISS2020), Marrakech, Morocco, pp. 1–8 (2020). https://doi.org/10.1145/3386723.3387844

29. Patel, R., Passi, K.: Sentiment analysis on twitter data of world cup soccer tournament using machine learning. IoT **1**(2), 218–239 (2020). https://doi.org/10.3390/iot1020014

30. Wang, Y., Chen, Q., Shen, J., Hou, B., Ahmed, M., Li, Z.: Aspect-level sentiment analysis based on gradual machine learning. Knowl. Based Syst. **212**, 106509–106521 (2021). https://doi.org/10.1016/j.knosys.2020.106509

31. Es-sabery, F., Hair, A.: A MapReduce C4.5 decision tree algorithm based on fuzzy rule-based system. Fuzzy Inf. Eng. 1–28 (2020). https://doi.org/10.1080/16168658.2020.1756099

32. Es-Sabery, F., Hair, A.: Big data solutions proposed for cluster computing systems challenges: a survey. In: Proceedings of the 3rd International Conference on Networking, Information Systems & Security, Marrakech, Morocco, pp. 1–7 (2020). https://doi.org/10.1145/3386723.3387826

33. Hamzi, B., Owhadi, H.: Learning dynamical systems from data: a simple cross-validation perspective, part I: parametric kernel flows. Physica D Nonlinear Phenom. **421** (2021). https://doi.org/10.1016/j.physd.2020.132817

A Convolutional Neural Networks-Based Approach for Potato Disease Classification

Khalid El Moutaouakil$^{(\boxtimes)}$ ⓘ, Brahim Jabir ⓘ, and Noureddine Falih

LIMATI Laboratory, Polydisciplinary Faculty, University of Sultan Moulay Slimane, Mghila,
BP 592 Beni Mellal, Morocco
`elmoutaouakil.kh@gmail.com`

Abstract. Identifying the various diseases that affect potato plants is a funda-mental thing for farmers to avoid losses every year. Among them, there are two common diseases known as early blight and late blight. The early detection of these diseases then the application of the appropriate treatment can prevent eco-nomic loss and save a lot of waste. The treatments for early blight and late blight are different thus it's important that we should accurately identify what kind of disease is in every potato plant. In this work, using deep learning, we propose a model that seeks to classify these potato plant diseases based on convolutional neural networks using one of the most widely used datasets. The model should be a solution that enables farmers to identify the early blight and late blight diseases present in their potato plants thus they can choose the appropriate treatment.

Keywords: Precision agriculture · Potato disease classification · Potato leaf blight · Convolutional neural networks · Deep learning

1 Introduction

Potato cultivation stands out in several countries in the world as one of the food products with the greatest impact on agricultural activities, given that, around its exploitation, the development of various sectors of the economy is generated [1], such as transport, industry, agrochemical distributors, packaging production, among others. One of the most common problems of this crop is the presence of early blight or late blight, affecting the leaves of the potato plant. Symptoms vary depending on various variables, like the climatic condition, among others. As example, on leaves, late blight begins as small light green spots that grow rapidly, turning grayish-brown.

Early diagnosis of plant diseases has great importance for sustainable and smart agriculture [2], to prevent unnecessary waste of financial resources, avoid the use of considerable amounts of pesticides and fungicides and to improve the quality of crops. The problem is that sometimes the plants do not show visible symptoms on their leaves; however, most of the time, diseases generate visible symptoms [3]. Thus, farmers need solutions to accurately identify the diseases. To improve the conditions that affect corps, precision agriculture is one of the main areas of research, which includes machine learning (ML) algorithms.

© Springer Nature Switzerland AG 2022
M. Fakir et al. (Eds.): CBI 2022, LNBIP 449, pp. 29–40, 2022.
https://doi.org/10.1007/978-3-031-06458-6_2

In recent years, computer vision, machine learning, and deep learning (DL) have become increasingly important due to their ability to process complex data with high accuracy. These techniques are the main ways to develop fast, automatic and accurate systems for image identification and classification. Currently, these systems are being implemented to automatically diagnose a wide variety of diseases present in various crops [4].

Convolutional neural networks (CNN) are one of the types of deep neural networks that have been well studied and have good performance in the detection and diagnosis of diseases, obtaining generic characteristics of data sets made up of images. However, there are challenges in implementing CNN networks. One of them is data collection for training, since a fairly large data set is required, containing a wide variety of conditions to work correctly. Therefore, part of the scientific community has opted for the use of publicly available data sets, among others, such as PlantVillage, which is a data set consisting of over 60,000 expertly curated images of healthy and infected leaves of plants.

2 Research Study

2.1 Deep Learning

Deep learning is a field of research that deals with finding theories and algorithms that allow a machine to learn on its own by stimulating neurons in the human brain. It is one of the branches of science that deals with artificial intelligence. In deep learning, we depend on Feed-forward neural network architectures that have multiple layers. The machine learns from big data using various designs of deep learning networks [5].

In general, we can say that each neural network is arranged in the form of layers of artificial cells: an inner layer, an outer layer, and layers between them or hidden between the input layers and the outer layer. Each neuron in one of these layers communicates with all the neurons in the next layer and all the neurons in the layer before it.

The output of a given neuron is as follows in Eq. (1):

$$Y = \sigma(w.x + b) \tag{1}$$

$x = (x1, x2, x3..., xn)$ is the input vector for a given neuron, $w = (w1, w2, w3...,wn)$ is the weight vector, b is the bias and σ represents one of the activation functions.

2.2 Convolutional Neural Networks

CNNs are a special type of feed-forward neural networks and is considered a solution to many computer vision problems in artificial intelligence, such as image classification, object detection within images and videos, face detection among others [6].

The training stage of the neural network constitutes a basic and important stage that comes after the stage of its design, as it is trained on a set of examples to give the network correct results on all the examples we have trained on.

2.3 Activation Functions

At the output of an artificial neuron, we apply an activation function which is a mathematical function that we should correctly choose to extract the maximum number of features [7, 8]. This choice is critical to make the neural network learn in adequate matter. There are several common activation functions, here are two examples used in our model:

Relu
We mostly use ReLU (Rectified Linear Unit) as an activation function in the context of artificial neural networks for its computational simplicity. For a given input x to a neuron, it returns 0 if x has a negative value, and the value of x in case it's positive.

$$f(x) = max(0, x) \tag{2}$$

Softmax
We use Softmax activation function σ in the last fully connected layer to scale numbers into probabilities, and we calculate it as follows:

$$\sigma\left(\overrightarrow{z}\right)_i = \frac{e^{z_i}}{\sum_{j=1}^{k} e^{z_j}} \tag{3}$$

$\left(\overrightarrow{z}\right)$ is the input vector
k is the number of classes in the multi-class classifier
e^{z_i} is the standard exponential function for the input vector
e^{z_j} is the standard exponential function for the output vector

3 Research Method

3.1 Hardware Configuration

The CNN architecture has been implemented using pandas, numpy, tensorflow and scikit-learn libraries, which use the Python programming language. The training and validation were carried out in Jupyter Notebook development environment using Python 3.9.7, while the model was tested on an HP laptop, with a Core i7 10th gen, 8 GB of RAM, 256 SSD and Windows 10 Pro operating system.

3.2 Dataset

The PlantVillage dataset is one of the most widely used in the world to build DL models in the agriculture field [9], in academic contexts. This dataset is 1.67 GB in size and features 39 different classes of healthy and infected leaves of 14 different crops, including apple, blackberry, corn, grape, peach, tomato, and potato. The images in this dataset are in the RGB color model, with a dimension of 256×256 pixels. For this study, the part of the PlantVillage dataset corresponding to the potato crop was used (Fig. 1).

Fig. 1. Sample images from the PlantVillage dataset

The used dataset consists of 3 image folders, 1,000 images correspond to the early blight classe, 1,000 images to the late blight and 152 images to the healthy. The images belong to crop samples from different areas of the world (Table 1).

Table 1. PlantVillage potato dataset.

PlantVillage potato dataset	
Class labels	Images
Early Blight	1000
Late Blight	1000
Healthy	152
Total images	**2152**

3.3 Data Augmentation

We propose a solution to classify the potato plant disease, and for this we based on the potato leaf disease PlantVillage dataset and we used data augmentation techniques [10] to increase the amount of data then we built a model using convolutional neural networks.

To increase the number of images in the PlantVillage dataset, data augmentation techniques are used to increase the healthy class of the data set by 760 additional samples and the infected class by 10000. The data augmentation techniques used in our model are: image flipping (horizontal and vertical flipping) and random rotation (Fig. 2).

Fig. 2. Leaf image from the augmented dataset flipped and rotated

The data augmentation process aims to increase the number of data and helps to avoid overfitting when training ML algorithms; this problem occurs when the network learns from particular cases instead of the general pattern.

3.4 Data Preparation

After applying data augmentation techniques, we divided the dataset into three subsets: training (80%), validation (10%) and test (10%). Table 2 shows the number of images used to feed our model after data augmentation. The training and validation datasets were only used to train the model, while the test set was in use to evaluate the performance with samples that the model did not consider before. In the training, features of the different types of potato crops that have the same symptoms were detected and, through evaluation metrics, the behavior of the architectures is estimated.

Table 2. Summary of the PlantVillage dataset after data augmentation.

PlantVillage potato dataset after data augmentation	
Class labels	Images
Early Blight	6000
Late Blight	6000
Healthy	912
Total images	**12912**

3.5 Model Architecture

The initial layer of our model is set for resizing the input images. Whenever the model gets and image, it will resize it and rescale it. The next layer is a convolution layer with filters of a (3 × 3) size, and we use ReLU as an activation layer, then we have a max pooling layer of size 2D. After this, we apply a set of convolution and max pooling layers all along the feature extraction phase. Regarding the classification phase, we flatten our pooled feature map to get a dense layer of 64 neurons then we apply Softmax activation function to normalize the probability of our classes (Fig. 3).

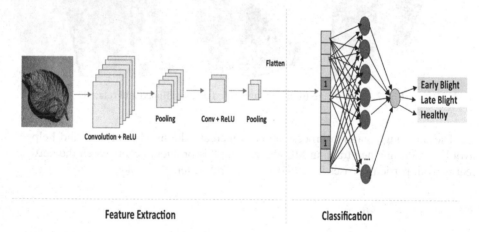

Fig. 3. Overall overview of the proposed model

The overall proposed architecture depends on 15 layers, including six convolutional layers and six max pooling layers. Adam is used as the optimizer, Cross entropy function for losses, accuracy as an evaluation metric and Softmax as the final judgment function. The total trainable parameters in use are 183,747.

Here is an overview of the model's architecture and the activation functions used in each layer (Fig. 4):

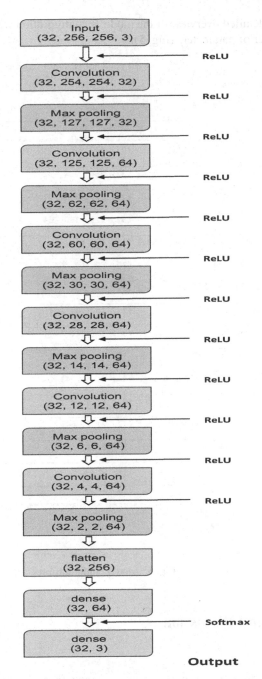

Fig. 4. Model architecture

Here is a detailed overview of the model's configuration with the size of each layer and the number of parameters (Fig. 5):

```
model.summary()

Model: "sequential_2"

Layer (type)                    Output Shape            Param #
=================================================================
sequential (Sequential)         (32, 256, 256, 3)       0

conv2d (Conv2D)                 (32, 254, 254, 32)      896

max_pooling2d (MaxPooling2D)    (32, 127, 127, 32)      0

conv2d_1 (Conv2D)               (32, 125, 125, 64)      18496

max_pooling2d_1 (MaxPooling2     (32, 62, 62, 64)       0

conv2d_2 (Conv2D)               (32, 60, 60, 64)        36928

max_pooling2d_2 (MaxPooling2     (32, 30, 30, 64)       0

conv2d_3 (Conv2D)               (32, 28, 28, 64)        36928

max_pooling2d_3 (MaxPooling2     (32, 14, 14, 64)       0

conv2d_4 (Conv2D)               (32, 12, 12, 64)        36928

max_pooling2d_4 (MaxPooling2     (32, 6, 6, 64)         0

conv2d_5 (Conv2D)               (32, 4, 4, 64)          36928

max_pooling2d_5 (MaxPooling2     (32, 2, 2, 64)         0

flatten (Flatten)               (32, 256)               0

dense (Dense)                   (32, 64)                16448

dense_1 (Dense)                 (32, 3)                 195
=================================================================
Total params: 183,747
Trainable params: 183,747
Non-trainable params: 0
```

Fig. 5. Model configuration

4   Results and Discussion

4.1   Results

We achieved an overall classification accuracy of 99.48%, and 0.63% of loss. The number of epochs selected is 54. Using the hardware configuration cited earlier, the training took 86 min and 30 s to finish. Our model, when compared with other previous related models presents a better performance according to the evaluation metrics (Fig. 6).

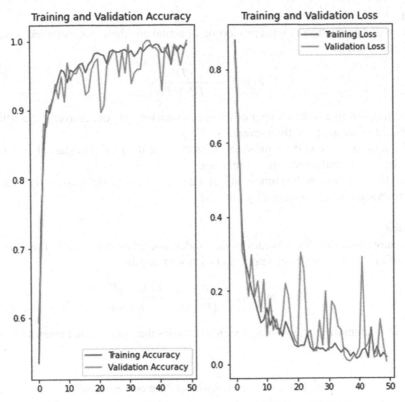

Fig. 6. Accuracy and loss evolution during training and validation

We applied several common performance evaluation metrics to fully evaluate the performance of the proposed architecture such as Accuracy, Precision, Recall and F1-score. The calculations for these metrics are presented in Eqs. (4), (5), (6) and (7).

4.2 Evaluation Metrics

Accuracy
Classification accuracy is a metric that gives the percentage of correct predictions of the model. We get it by dividing the number of correct predictions by the total number of predictions.

$$Accuracy = \frac{Number\ of\ Correct\ Predictions}{Total\ Number\ of\ Predictions} \tag{4}$$

Precision
Precision is a metric that gives the percentage of what proportion of positive identifications was actually correct. Mathematically, Precision is defined as:

$$Precision = \frac{TP}{(TP + FP)} \tag{5}$$

Recall

Recall gives a percentage of what proportion of actual positives was identified correctly. Recall is defined as:

$$Recall = \frac{TP}{(TP + FN)} \tag{6}$$

TP: The true positive is the number of instances which the model correctly identified as relevant, and they are actually relevant.

FP: The false positive is the number of instances that the model mistakenly identified as relevant, but actually they are not relevant.

FN: The false negative is the number of instances which the model incorrectly identified as not relevant, but they are actually relevant.

F1-Score

The F1-score measures the accuracy of the models and takes into account Precision and Recall of the test to classify examples as positive or negative.

$$F1 - score = 2x \frac{Precision\,x\,Recall}{(Precision + Recall)} \tag{7}$$

Table 3 summarizes the values of various metrics that we obtained when testing our model.

Table 3. Classification metrics of the proposed model.

Metric	Value
Accuracy	99.48%
Precision	99.56%
Recall	99.98%
F1-score	99.77%

4.3 Comparison with Related Works

In comparison with previous related works, when applying the same data augmentation techniques, our model provides higher accuracy, as you can see in Table 4 below:

Table 4. Comparison with related works.

Related works			
Methodology	Accuracy	Year	Reference
CNN	92.00%	2020	[11]
FFNN	96.50%	2020	[12]
SBCNN	96.75%	2020	[13]
SVM, KNN and Neural Net	97.80%	2020	[14]
CNN	98.00%	2021	[15]
CNN	98.28%	2021	[16]
CNN	99.00%	2020	[17]
CNN (Our model)	**99.48%**	**2022**	

5 Conclusion and Future Work

Early blight and late blight diseases in potato plants constitute a real problem for farmers and threaten food security in all nations, thus the early detection of these diseases is important. For this, we based on convolutional neural networks, which gives the opportunity to develop solutions for this kind of problems in precision agriculture. This paper presents a deep learning model capable of recognizing potato early blight and late blight by classifying leaf images. In the application of this model, the augmented data set of PlantVillage was used for training. The proposed model has been evaluated based on performance metrics, such as precision, F1 score and recall. The experimental results obtained with the selected data set showed that the proposed model reaches an accuracy of 99.48%. Our architecture achieved better results than the related previous works in terms of overall accuracy. Therefore, it is concluded that the model is a valid solution for farmers in the identification of early blight and late blight. One prospect for future work is to reduce the computation time in the training phase using pre-trained architectures like AlexNet, VGG16, and VGG19.

References

1. Pretty, J.: Agricultural sustainability: concepts, principles and evidence. Philos. Trans. R. Soc. B: Biol. Sci. **363**(1491), 447–465 (2008)
2. Tilman, D., Cassman, K.G., Matson, P.A., Naylor, R., Polasky, S.: Agricultural sustainability and intensive production practices. Nature **418**(6898), 671–677 (2002)
3. Gavhale, K.R., Gawande, U.: An overview of the research on plant leaves disease detection using image processing techniques. IOSR J. Comput. Eng. IOSR-JCE **16**(1), 10–16 (2014)
4. Jabir, B., Falih, N.: Digital agriculture in Morocco, opportunities and challenges. In: The 6th International Conference on Optimization and Applications, pp. 1–5 (2020)
5. Liu, W., Wang, Z., Liu, X., Zeng, N., Liu, Y., Alsaadi, F.E.: A survey of deep neural network architectures and their applications. Neurocomputing **234**, 11–26 (2017)

6. Jabir, B., Falih, N., Rahmani, K.: Accuracy and efficiency comparison of object detection open-source models. Int. J. Online Biomed. Eng. **17**(5) (2021)
7. Krizhevsky, A., Sutskever, I., Hinton, G.E.: ImageNet classification with deep convolutional neural networks. In: Advances in Neural Information Processing Systems, vol. 25 (2012)
8. Singh, R.G., Kishore, N.: The impact of transformation function on the classification ability of complex valued extreme learning machines. In: 2013 International Conference on Control, Computing, Communication and Materials (ICCCCM), pp. 1–5 (2013)
9. Mallah, C., Cope, J., Orwell, J.: Plant leaf classification using probabilistic integration of shape, texture and margin features. Signal Process.Pattern Recogn. Appl. **5**(1), 45–54 (2013)
10. Xie, Q., Dai, Z., Hovy, E., Luong, T., Le, Q.: Unsupervised data augmentation for consistency training. Adv. Neural. Inf. Process. Syst. **33**, 6256–6268 (2020)
11. Rozaqi, A.J., Sunyoto, A.: Identification of disease in potato leaves using Convolutional Neural Network (CNN) algorithm. In: 2020 3rd International Conference on Information and Communications Technology (ICOIACT), pp. 72–76 (2020)
12. Sanjeev, K., Gupta, N.K., Jeberson, W., Paswan, S.: Early prediction of potato leaf diseases using ANN classifier. Orient. J. Comput. Sci. Technol. **13**(2, 3), 129–134 (2021)
13. Barman, U., Sahu, D., Barman, G.G., Das, J.: Comparative assessment of deep learning to detect the leaf diseases of potato based on data augmentation. In: International Conference on Computational Performance Evaluation, pp. 682–687 (2020)
14. Tiwari, D., Ashish, M., Gangwar, N., Sharma, A., Patel, S., Bhardwaj, S.: Potato leaf diseases detection using deep learning. In: 2020 4th International Conference on Intelligent Computing and Control Systems (ICICCS), pp. 461–466 (2020)
15. Khalifa, N.E.M., Taha, M.H.N., Abou El-Maged, L.M., Hassanien, A.E.: Artificial intelligence in potato leaf disease classification: a deep learning approach. In: Hassanien, A.E., Darwish, A. (eds.) Machine Learning and Big Data Analytics Paradigms: Analysis, Applications and Challenges. SBD, vol. 77, pp. 63–79. Springer, Cham (2021). https://doi.org/10.1007/978-3-030-59338-4_4
16. Rashid, J., Khan, I., Ali, G., Almotiri, S.H., AlGhamdi, M.A., Masood, K.: Multi-level deep learning model for potato leaf disease recognition. Electronics **10**(17), 2064 (2021)
17. Lee, T.Y., Yu, J.Y., Chang, Y.C., Yang, J.M.: Health detection for potato leaf with convolutional neural network. In: The 2020 Indo–Taiwan 2nd International Conference on Computing, Analytics and Networks, pp. 289–293 (2020)

Performance Investigation of a Proposed CBIR Search Engine Using Deep Convolutional Neural Networks

Smail Zitan[1]([✉]) [ID], Imad Zeroual[2] [ID], and Said Agoujil[1] [ID]

[1] MMIS, MAIS, FST Errachidia, Moulay Ismail University, Meknes, Morocco
zitansmail@gmail.com
[2] L-STI, T-IDMS, FST Errachidia, Moulay Ismail University, Meknes, Morocco

Abstract. During the last decade, the volume of image databases and collections has significantly increased. Further, most current image retrieval systems are moving from traditional methods, which are mainly metadata-based approaches, to Content-Based Image Retrieval (CBIR) approaches. These latter use an image as an input query without considering any metadata associated with the image. Generally, a CBIR-based system extracts visual features such as color, image edge, and texture to search and retrieve similar images. Our goal is to develop a fast and accurate CBIR-based search engine. In this paper, we investigated the performance of two Convolutional Neural Networks (CNN) models, namely VGG16 and MobileNet, in a real-life project environment. Further, we used both Faiss and Annoy libraries for indexing and similarity search, while the Principal Component Analysis (PCA) technique is applied for reducing the dimensionality of the feature vector space. From the experiments, we found that the VGG16 and MobileNet models differ in terms of precision, recall, and response time recorded. However, they both have succeeded in supporting the CBIR-based search engine.

Keywords: CBIR search engine · Convolutional Neural Networks · VGG16 · MobileNet · Faiss · Annoy · Principal Component Analysis

1 Introduction

Since its appearance, the search for a special object in the World Wilde Web application was often long and complicated, especially with the increase of data. The traditional search for images primarily uses metadata-based approaches, which is computationally expensive and relatively inaccurate.

In the decade, images search engines (e.g., Google and Bing) were moving toward using Content-Based Image Retrieval (CBIR) approaches that allow the users to find a similar set of images based on input in the form of images instead of using just keywords. Such CBIR-based search engines might replace textual queries and make it easier for users to do quick searches with minimal information.

In this paper, we investigated the performance of a proposed CBIR-based search engine using two different deep Convolutional Neural Networks, namely VGG16 and

© Springer Nature Switzerland AG 2022
M. Fakir et al. (Eds.): CBI 2022, LNBIP 449, pp. 41–49, 2022.
https://doi.org/10.1007/978-3-031-06458-6_3

MobileNet. The performance of these two models is evaluated using two different search indexing algorithms, namely, Annoy and Faiss. To handle the high dimensionality of data, we applied the Principal Analysis Component (PCA) method. A freely available dataset with 28k animal images, which is classified into 10 categories, is used. Finally, the results are reported using several standard metrics such as precision, recall, and response time.

In addition to the introduction, the rest of this paper is arranged in the following four sections. In Sect. 2, we listed the relevant related works that implemented CNN models for CBIR-based systems. In Sect. 3, we described the dataset used in this evaluation. Further, we presented the methodology adopted. In Sect. 4, we reported the results obtained in the experiments conducted alongside a discussion. In Sect. 5, we concluded the paper and provided some ideas for future works.

2   Related Works

Content-Based Image Retrieval (CBIR) approaches search a group of images that have similarities with the input images. The CBIR-based system extracts visual features, then, it performs a classification task during the retrieval process to retrieve the images that belong to the same class as the input image.

An earlier study [1] have used a CNN architecture called AlexNet [2] as a feature extraction method. The authors used both the ILSVRC-2010 and ILSVRC-2010 datasets that comprises about 1000 images in each of 1000 categories. These datasets are subsets of the ImageNet dataset [3] which includes more than 15 million images manually classified into 22,000 categories using the crowdsourcing tool Mechanical Turk of Amazon. The proposed model is based on a deep convolutional neural network with stochastic gradient descent. The reported error rates are 17.0% and 15.3% for ILSVRC-2010 and ILSVRC-2012, respectively.

A following study [4] have developed a CBIR-based system using the VGG16 [5] as a visual feature extraction method, while the retrieval process is performed using the cosine similarity [6]. The authors used a subset of the iNaturalist dataset [7]. The original iNaturalist dataset comprises 13 masterclass, 5089 subclasses, and 579184 training images. However, the experiment involved only five masterclass and each of them has five subclasses. Four of masterclass are animals, while one is Plantae. In order to test to retrieval model, the authors used images from the Internet (i.e., not included in the iNaturalist dataset). Each subclass was tested twice with 2 different images. The retrieved images from the masterclass contain 250 images. The CBIR-based system achieved an average precision of 89.6% and 17.92% for recall (maximum value for the recall was 20%).

A more recent study by Desai et al. [8] also used the VGG16 as a visual feature extraction method; however, the authors used the Support Vector Machine (SVM) algorithm [9] for images classification. In the end, the retrieval method is based on similarity index that is the measure of the distance values with respect to the query image. In their experiments, the authors used the COREL-1k dataset [10]. This latter has 1,000 images divided equally on 10 categories, namely African tribes, Beaches, Buildings, Buses, Dinosaurs, Elephants, Flowers, Horses, Mountains, and Food. Regarding the query set, it consists of randomly selected 10 images. The proposed model achieved an average precision of 83.5%.

3   Materials and Methods

3.1   Dataset

The dataset used in this study is publicly available on Kaggle[1]. This dataset has 585MB size, it contains about 28K medium quality animal images. The images were classified into 10 categories, namely dog, cat, horse, spider, butterfly, chicken, sheep, cow, squirrel, and elephant. The number of images per category varies from 2K to 5K units. All the images have been compiled using "google images" and have been manually checked. Figure 1 exhibits the number of images per category.

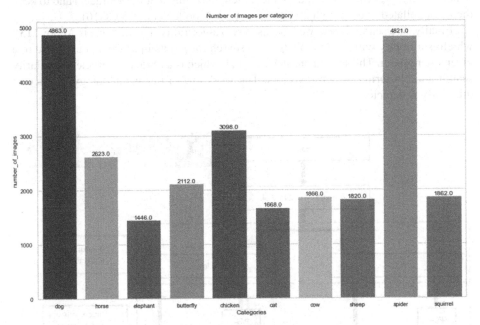

Fig. 1. Number of images per category

3.2   Methods

Our proposed CBIR-based search engine consists of four main phases. The first one is the pre-processing phase when images are reshaped to new dimension 224 × 224. Then, visual feature extraction is applied using CNN models. Dimensionality reduction is performed in the third phase. Finally, the retrieval process is implemented using similarity search methods.

For the purpose of comparison, we have used two CNN models for extracting features from the used dataset. The first CNN model is VGG16 [5] and the second one is MobileNet [11]. Both are originally trained on over 1M images from the ImageNet dataset [3]. VGG16 is composed of 13 convolutional layers, 5 max-pooling layers, and 3

---

[1] https://www.kaggle.com/alessiocorrado99/animals10.

fully connected layers. Therefore, there are 16 layers in total having tunable parameters (i.e., 13 convolutional layers and 3 fully connected layers). In this study, we only performed the VGG16 as a feature extraction method, therefore, we removed the last three fully connected layers that perform the classification task in the VGG16 architecture.

Regarding the MobileNet architecture, it scales the input size of the image on 224 × 224 images with a global pooling. The fully connected classifier at the end of the network depends only on the number of channels, and not the feature maps spatial dimension. Similarly, we removed the fully connected layers since we only want to extract the features from images.

As for dimensionality reduction phase, we have applied the Principal Analysis Component (PCA) [12] method to reduce the dimension of the features extracted and to keep the most collated ones in both architectures (i.e., MobileNet and VGG16).

Finally, as a search index, we have used two algorithms. The first one is Annoy [13], which is a library with Python bindings to search for points in space that are close to a given query point. The second one is Faiss [14], which is a library for efficient similarity search and clustering of dense vectors. The workflow of our methodology adopted in this study is depicted in Fig. 2.

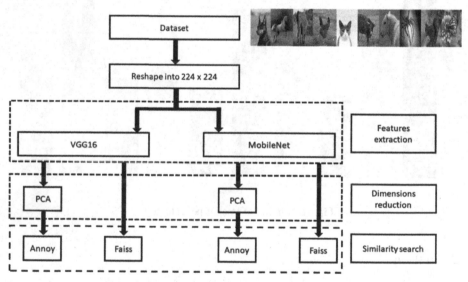

Fig. 2. Workflow of the proposed CBIR-based search engine.

To sum up, we implemented two approaches to build a CBIR search engine based on two different CNN models with two distinct search algorithms. Both CNN models (i.e., VGG16 and MobileNet) are used for the feature extraction phase, while Faiss and Annoy are applied as search indexing algorithms. Finally, the well-known PCA is used as a feature dimensionality reduction method. To the best of our knowledge, these two approaches are still not yet implemented and compared in a CBIR search engine.

3.3 Pre-processing

The dataset comprises about 28K images. We conducted a pre-processing task to standardize the size of images. For deep learning-based CNN approach, we resized all images in the dataset into 224 × 224 pixels.

3.4 Deep Learning-Based CNN

Deep learning is becoming increasingly popular for extracting information from images. We used a CNN in this experiment with two separate pre-trained models, VGG16 and MobileNet. CNN is a sort of multilayer neural networks that have a grid-like architecture and is used to process an image as an input data. It can be thought of as a 2D grid of pixels, then, image pixels can be analyzed directly using the CNN to recognize image patterns.

The input layer, hidden layer, and output layer are the three layers that make up the CNN architecture, the task of constructing CNN architecture is crucial since it has an impact on image recognition results.

As a result, various CNN architectures for image identification, such as VGG16 and MobileNet, have been introduced. In this experiment, we resize the input pixel size of VGG16 and MobileNet. Figure 3 summarizes the information of VGG16 architectures, while, the architecture of MobileNet is shown in Fig. 4.

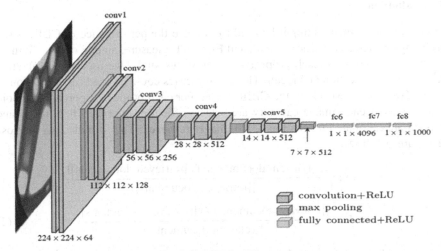

Fig. 3. VGG16 architecture

After the pre-processing phase, where the input image is resized into a fixed-size 224 × 224 pixels, the input image is passed via a pile of convolutional layers (i.e., conv1, conv2, conv3, conv4, conv5) and five max-pooling layers. These latter are considered as filters where the Kernel extracts the maximum value of the area it convolves. Finally, the three fully connected layers perform the classification task, which is not applied in our case.

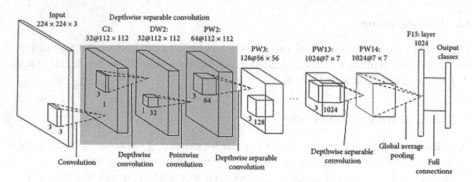

Fig. 4. MobileNet architecture.

As Fig. 4 depicted, MobileNet architecture employs several depthwise separable convolutions. The name separable originated from the idea that a filter's depth and spatial dimension can be separated. Consequently, each depthwise separable convolution operates a 3 × 3 depthwise convolution followed by a 1 × 1 pointwise convolution. The global average pooling is a pooling operation designed to generate one feature map for each corresponding category. The final fully connected layers are used for the classification task; however, we did not involve them in our approach.

3.5 Evaluation

Several evaluation metrics might be used to evaluate the performance of CBIR-based search engine such as precision, recall, and F-score/F-measure, among others. To measure the effectiveness of each proposed model in this study, we used two evaluation metrics namely precision and recall. These two metrics could be used for evaluation of any retrieval model, especially the CBIR model. For each proposed approach, random 10 images from each image category are selected for user queries. Then, recall and precision are calculated for the top 1,000 images retrieved. The formulations of those metrics are as follows:

$$\text{precision} = \frac{|\{\text{relevant documents}\} \cap \{\text{retrieved documents}\}|}{|\{\text{retrieved documents}\}|} \tag{1}$$

$$\text{recall} = \frac{|\{\text{relevant documents}\} \cap \{\text{retrieved documents}\}|}{|\{\text{relevant documents}\}|} \tag{2}$$

Additionally, we considered the mean of response time after making the request 10 times as rapidity for each approach:

$$\text{rapidity} = \frac{\sum_{k=1}^{10} q_k}{10} \tag{3}$$

4   Results and Discussion

The aim of this experiment is to compare the performance of different approaches for image search engine that give the high precision with minimum of response time. Table 1 presents the precision, recall, and rapidity over images in all 10 categories using the VGG16 model with both search index Annoy and Faiss.

Table 1. VGG16 result with annoy and Faiss.

Class	VGG16					
	Annoy			Faiss		
	Precision (%)	Recall (%)	Rapidity (s)	Precision (%)	Recall (%)	Rapidity (s)
Dog	53	11	1.6	41	8.43	2.74
Cavallo	49.9	19	1.05	57.7	21.9	2.23
Elephant	99.6	68.8	1.02	96.7	66.8	2.24
Papillon	81.9	38.7	1.04	54.4	25.7	2.28
Chicken	100	32.2	1.08	99.8	32.2	2.32
Cat	87.8	52.6	1.15	81.2	48.6	2.37
Cow	44.7	23.9	1.06	48.7	26	2.27
Sheep	93.1	51.1	1.05	73.5	40.3	2.27
Spider	100	20.74	1.11	100	20	2.20
Squirrel	100	53.7	1.10	99.3	53.3	2.23
Average	81	37.17	1.12	75.23	34.32	2.31

As shown in Table 1, applying VGG16 with Annoy achieved better results in terms of precision (81%), recall (37.17%) and response time (1.12 s) compared to VGG16 with Faiss, which records average rates of 75.23%, 34.32%, and 2.31s for precision, recall, and response time, respectively.

Similarly, Table 2 illustrates the precision, recall, and rapidity over images in all 10 categories using the MobileNet model with both search index Annoy and Faiss.

Again, the Annoy algorithm yields to better performance than Faiss while using the MobileNet model in all different performance metrics. Precisely, the MobileNet model achieved average rates of 70.23%, 32.06%, and 0.31s for precision, recall, and response time, respectively. Whereas, using Faiss yields to average rates of 67.03%, 31.31%, and 0.61s for precision, recall, and response time, respectively.

As for the comparison between the VGG16 and MobileNet models, the results obtained by the VGG16 architecture are higher than those recorded by MobileNet. The VGG16 with Annoy scored 81% and 37.17% for precision and recall, respectively. Whereas, the MobileNet with Annoy achieved 70.23% and 32.06% for precision and recall, respectively. Similarly, the VGG16 with Faiss obtained 75.23% and 34.32% for precision and recall, respectively. In contrast, the MobileNet with Annoy

Table 2. MobileNet result with annoy and Faiss.

| Class | MobileNet | | | | | |
| | Annoy | | | Faiss | | |
	Precision (%)	Recall (%)	Rapidity (s)	Precision (%)	Recall (%)	Rapidity (s)
Dog	49.5	10.17	0.32	42.6	9.5	0.62
Cavallo	56.1	21.3	0.31	62.6	23.8	0.59
Elephant	92.1	63.6	0.29	80.9	66.8	0.56
Papillon	90.7	42,9	0.32	74.4	35.2	0.62
Chicken	58.3	18.8	0.31	67.2	21.6	0.63
Cat	79.2	47.4	0.37	65.2	39	0.66
Cow	33.5	17.9	0.30	40.9	21.9	0.61
Sheep	80.7	44.3	0.31	70.6	38.79	0.65
Spider	99.6	20.65	0.30	98.8	20.49	0.60
Squirrel	62.6	33.61	0.31	67.1	36.03	0.63
Average	70.23	32.06	0.31	67.03	31.31	0.61

recorded 67.03% and 31.31% for precision and recall, respectively. On the other hand, the MobileNet is faster than the VGG16 by an average factor of 3.72.

These results can be explained by the fact that the architecture of VGG16 has more trainable parameters which leads to extract more deep features from the input images. Consequently, the deep features extracted by VGG16 improved the CBIR performance to obtain higher precision and recall, while negatively impacting its response time. On the contrary, the MobileNet architecture has extracted fewer features which explains its capability of giving results quicker while the precision and recall are relatively inferior.

5   Conclusions and Perspectives

In this study, we have evaluated two different CNN architectures (i.e., VGG16 and MobileNet) alongside two different search indexes (i.e., Annoy and Faiss) for CBIR-based search engine. VGG16 and MobileNet architectures are deep learning-based CNN that were previously trained models on a part of the ImageNet dataset. Both models used similar pre-processed images with 224 × 224 dimension and the PCA as a dimensionality reduction method.

From the experiments conducted, the results obtained show that using Annoy as search indexing algorithm yields to better results in terms of precision, recall, and response time with both CNN models compared to Faiss. Further, The VGG16 outperforms the MobileNet in terms of precision and recall; whereas, in term of response time, the MobileNet is faster than VGG16,

For future works, further improvements are still at hand if we consider using other feature extraction techniques and search indexing algorithms. The effects of alternative methods to reduce the dimensions of extracted features can be tested. Also, we plan to perform more investigations using different datasets, while implementing other approaches existed in the literature.

References

1. Huang, W., Wu, Q.: Image retrieval algorithm based on convolutional neural network. In: Current Trends in Computer Science and Mechanical Automation, Sciendo Migration, vol. 1, pp. 304–314 (2017)
2. Krizhevsky, A., Sutskever, I., Hinton, G.E.: ImageNet classification with deep convolutional neural networks. Commun. ACM **60**, 84–90 (2017). https://doi.org/10.1145/3065386
3. Deng, J., Dong, W., Socher, R., Li, L.-J., Li, K., Fei-Fei, L.: Imagenet: a large-scale hierarchical image database. In: 2009 IEEE Conference on Computer Vision and Pattern Recognition, pp. 248–255. IEEE (2009)
4. Rian, Z., Christanti, V., Hendryli, J.: Content-based image retrieval using convolutional neural networks. In: 2019 IEEE International Conference on Signals and Systems (ICSigSys), pp. 1–7. IEEE (2019)
5. Simonyan, K., Zisserman, A.: Very deep convolutional networks for large-scale image recognition. arXiv preprint arXiv:1409.1556 (2014)
6. Huang, A.: Similarity measures for text document clustering. In: Proceedings of the Sixth New Zealand Computer Science Research Student Conference (NZCSRSC2008), Christchurch, New Zealand, pp. 9–56 (2008)
7. Van Horn, G., et al.: The inaturalist species classification and detection dataset. In: Proceedings of the IEEE Conference on Computer Vision and Pattern Recognition, pp. 8769–8778 (2018)
8. Desai, P., Pujari, J., Sujatha, C., Kamble, A., Kambli, A.: Hybrid approach for content-based image retrieval using VGG16 layered architecture and SVM: an application of deep learning. SN Comput. Sci. **2**(3), 1–9 (2021). https://doi.org/10.1007/s42979-021-00529-4
9. Vapnik, V., Chervonenkis, A.Y.: A class of algorithms for pattern recognition learning. Avtomat. i Telemekh. **25**, 937–945 (1964)
10. Li, J., Wang, J.Z.: Real-time computerized annotation of pictures (2011)
11. Howard, A.G., et al.: Mobilenets: efficient convolutional neural networks for mobile vision applications. arXiv preprint arXiv:1704.04861 (2017)
12. Dunteman, G.H.: Principal Components Analysis. Sage, Thousand Oaks (1989)
13. Bernhardsson, E.: Annoy: approximate nearest neighbors in C++/python. Python Package Version 1 (2018)
14. Johnson, J., Douze, M., Jégou, H.: Billion-scale similarity search with GPUs. IEEE Trans. Big Data (2019)

Decision Boundary to Improve the Sensitivity of Deep Neural Networks Models

Mohamed Ouriha[1]([⊠]) [iD], Youssef El Habouz[2], and Omar El Mansouri[1] [iD]

[1] TIAD Laboratory, Sciences and Technology Faculty, Sultan Moulay Slimane University,
Beni Mellal, Morocco
mohamed.ouriha@usms.ma
[2] IGDR UMR 6290 CNRS Rennes 1 University, Rennes, France

Abstract. In spite of their performance and relevance on various image classification fields, deep neural network classifiers encounter real difficulties face of minor information perturbations. In particular, the presence of contradictory examples causes a big weakness and insufficiency of deep learning models in many areas, such as illness recognition. The aim of our paper is to improve the robustness of deep neural network models to small input perturbations using standard training and adversarial training to maximize the distance between predict instances and the boundary decision area. We shows the decision boundary performance of deep neural networks during model training, the minimum distance of the input images from the decision boundary area and how this distance develops during the deep neural network training. The results shows that the distance between the images and the decision boundary decreases during standard training. However, adversarial training increases this distance, which improve the performance of our model. Our work presents a new solution to the deep neural networks sensitivity problem. We found a very strong relationship between the efficiency of the deep neural networks model and the training phase. We can say that the efficiency is created during training, it is not predetermined by the initialization or architecture.

Keywords: Deep neural network · Boundary decision · Adversarial training · Image classification

1 Introduction

Deep learning has supported developments in a variety of fields. However, just because we have high testing performance doesn't mean we have a good understanding of how they work. In recent years, deep neural networks have been demonstrated to be exceedingly susceptible. In the face of minor input disruptions, they are particularly helpless [14, 18].

This insight arguably contradicts the normal belief that deep networks adapt well to similar, unseen examples. Such sources of dangerous information may be the result of distribution changes or a general disturbance in the classifier's environment [19]. For example, unusual lighting or weather conditions for an independent vehicle. Or they

M. Fakir et al. (Eds.): CBI 2022, LNBIP 449, pp. 50–60, 2022.
https://doi.org/10.1007/978-3-031-06458-6_4

can be mismatched patterns, such as regular disturbed images intentionally generated by an enemy to cause classification errors. Previously, these two types of destructive interference were mostly studied by separate research networks, namely dirt robustness specialists and enemy robustness specialists. Recently, it has become increasingly clear that the susceptibility to these different disturbances is tightly linked and therefore should not be studied independently [27].

For example, using the test error in additive Gaussian noise, Ford et al. [12] generated an estimate of the size of small adverse disturbances.

In general, we are still in the early stages of understanding the decision-making process and the obstacles that stand in the way of deep neural networks.

The course of the decision-making frontier and the factors affecting it require more detailed study because it can guide us to viable adversarial defenses.

In this paper, we present an empirical analysis that focuses on the distance between the image and the decision boundary and how this distance changes as the classifier is trained. At MNIST, our approach to neural network classifiers has led to three observations:

- The training and test images are close to the decision boundary. Even in the late stages of training, when the neural network has achieved low error rates, this phenomenon persists.
- In contrast to the first result, Adversarial training causes the decision boundary to evolve in a very different way. In comparison to normal training, the average distance between the images and the decision boundary stays greater. This study found that adversarial training has increased the durability of deep neural network models for classification tasks.
- The images that have been misclassified are closer to the decision boundary than the ones that have been successfully classified. The decision boundary is moved towards these images during training. This belief holds true for both adversarial and normal training.

2 Theoretical Background

In this part, we will define the concept of a decision boundary and examine the DeepFool method.

2.1 The Decision Boundary

Mickisch et al. [16] clarify the concept of decision boundary by defining the classifier as a function. $f : R^n \rightarrow R^c$ where n represents the number of variables in the input data (the number of pixels in a image) and c defines the number of classification categories. The output $f(x)$ for an input image $x \in R^n$ may be understood as the classifier's vector of softmax values, with the classification decision provided by $K(x) = \text{argmax}_{k=1,\dots,c} f_k(x)$ The decision boundary is specifically defined with this syntax:

$D \subseteq R^n$ of f as the set

$$D := \{x \in R^n | \ni k_1, k_2 = 1, \dots, c, k_1 \neq k_2, \\ f_{k1}(x) = f_{k2}(x) = \max_k f_k(x)\}. \tag{1}$$

To put it another way, these are the places at which the classifier's conclusion is in doubt. The data point x's margin $d_2(x) \in R_{\geq 0}$ is calculated as follows:

$$d_2(x) = \min_{\delta \in R^n} \|\delta\|_2$$
$$s.t. x + \delta \in D \qquad (2)$$

The l_2-norm is used in the preceding margin expression. During our empirical research, we'll additionally take into account the margin relative to the l_2-norm, that we'll name d_2.

2.2 DeepFool

DeepFool is an unfocused, repetitive, adversarial assault that terminates when $K(x_0 + \delta) \neq K(x_0)$ finds a perturbation for input data point x_0. DeepFool was developed by Moosavi-Dezfooli et al. [24] with the purpose of providing a technique that can efficiently estimate adversarial perturbations while also providing a better evaluation of the durability of f at x_0. The experts do this by using the well-known orthogonal projection of a point along with a linear classifier's decision boundary. Specifically, the classifier's class probability functions f_k $k = 1, \ldots, c$ are normalized around the present position x_i at each iteration stage of DeepFool. The minimal disturbance δ_i with regard to the -norm is then calculated, which shifts xi onto the linearized model of f's decision boundary. Every perturbation δ_i may be expressed as follows:

$$\delta_i := \frac{\left| f_l(x_i) - f_{k(x0)}(x_i) \right|}{\left\| \nabla f_l(x_i) - \nabla f_{k(x0)}(x_i) \right\|_2^2} \left(\nabla f_l(x_i) - \nabla f_{k(x0)}(x_i) \right) \qquad (3)$$

with index

$$l = l(x_i) := \operatorname{argmin}_{k \neq k(x0)} \frac{\left| f_l(x_i) - f_{k(x0)}(x_i) \right|}{\left\| \nabla f_l(x_i) - \nabla f_{k(x0)}(x_i) \right\|_2} \qquad (4)$$

For image x_0, N is the iteration's terminating index. i.e. $K(x_N) \neq K(x_0)$. The required adversarial perturbation is expressed as

$$\delta = \sum_{i=0}^{N-1} \delta_i \qquad (5)$$

In general, the DeepFool attack may be thought of as a function.

DeepFool:$R^n \to R^n$ It takes a image x_0 and generates an adversarial perturbation for the provided classifier f. The researchers of the original studies [24] also propose DeepFool algorithm modifications for every l_p-distance value for $p \in [1, \infty]$. We'll use the l_2-adjustment for these estimations because we want to analyze the margin in relation to the l_2-norm during our experimental analysis.

3 Related Work

The research on the decision boundary of deep learning models is presented in this section. There has been a substantial lot of study effort focused on researching techniques that produce adversarial instances [31, 34] and also strategies that try to develop robustness [25, 28] using the foundational work of Biggio et al. [33] and Szegedy et al. [20] as a beginning point. According to PGD [7] and C&W [15], adversarial assaults in machine vision may effectively generate any wanted classification result by introducing an adversarial disturbance to the natural image. Metzen et al. [6] proved this in the situation of semantic separation. In this situation, the researchers could use generalized adversarial disturbances to get any segmentation they wanted. As a consequence of these adversarial attacks, we can see that the decision boundary of a traditional deep neural network is close to any source image. In other terms, the decision boundary's closeness to the real image reveals the vulnerability of deep neural network models in the face of unique source perturbations. As a result, a large number of adversarial strategies attempt to raise the minimum distance between real images and the decision.

Furthermore, experts have developed specialized architectures [3, 23], normalization penalties [11, 32], and data augmentation approaches [2, 4] in order to get a more favorable trajectory of the decision boundary and, as a result, more durable classifiers. Unfortunately, most defenses have failed and have been destroyed by more solid opponents not long after their introduction [17].PGD adversarial training is still widely regarded as the most trustworthy and effective adversarial defense [7].

In every event, it's worth noting that adversarial training works best with extremely limited It has a tendency to overfit specific assaults into threat models. Rather than gaining general resilience [8]. Because of the slow progress in developing robustness, a growing group of experts are wondering if deep neural networks will ever achieve robustness [13, 21, 22].

The notable, but very speculative, argument that neural networks with a solitary hidden layer are global function estimation has mostly provided us with a misguided feeling of safety. Recent research, for instance, suggests that frequently utilised neural network architectures with a high lot of low-hidden layers cannot lead to global function estimation [26].

Furthermore, studies have demonstrated that the decision regions of recent machine learning classifiers, i.e., the areas of the input space that ending in a certain output class, have to be linked groups [26, 30]. This high topological limitation on the decision areas may already be limiting these models' expressive capability and, consequently, their maximum attainable resilience. Furthermore, there have been attempts to define universal durability constraints for classification issue categories that are independent of the classification algorithm utilized. Fawzi et al. [13] give basic maximum constraints on the resilience that may be achieved when the real data originates from a smooth, productive model. They illustrate that every sort of classifier is vulnerable to antagonist disturbances when the hidden space of the productive model is high dimensional, based on this assumption about the data's origin.

The existence of adversarial instances for best-in-class neural networks shows that natural images are very likely to be near to the decision boundary D, so $d_2(x)$. is small for the greater part of input x Deep neural networks, on the other hand, are notoriously resistant to noise [1]. In this method, the decision boundary might be reasoned to be near in a few directions while being further distant in the majority of perturbation directions. The authors of [10] confirm this impression by adding carelessly picked symmetrical directions to benign images that appear combative. These sampling perturbations seldom impact the category choice for good images. They find, however, that the large number of adversarial models created by the FGSM attack [29] are vulnerable to these arbitrary deviations.

We'd like to use an optimization approach to calculate $d_2(x)$ for our studies, but because the border optimization issue (2) is difficult for deep neural networks, we'll have to settle for an estimated answer. Specifically, we're seeking for a minor δ disturbance that results in $K(x + \delta) \neq K(x)$ Following that, ideally, have $d_2(x) \approx \delta_2$ however $x + \delta$ is not always a member of D [16].

With the help of DeepFool, we'll calculate the margin estimate. Other strong adversial assaults, on the other hand, can also be employed to create δ. The PGD attack, in particular, is used in [5] to estimate the distance between the training examples and the decision boundary. They discover that the standard cross-entropy loss is one reason for the tiny margins, and that a different training process produces more solid models. DeepFool, on the other hand, has been shown to create minor adversial disturbances, making it suitable for border approximation [16]. To calculate $d_2(x)$ for images from the training data, Jiang et al. [9] used a simplified version of DeepFool with a single iteration phase. Other than measuring the distance to the decision boundary, it's also useful to know the decision boundary's more generic structural characteristics.

4   Methodology

In this section, we define our proposed approach as well as the database and the model used.

4.1   Dataset

In our work, we use the MNIST dataset. MNIST is a collection of handwritten numbers. In deep learning, it is an extensively used dataset. There are 60,000 training pictures and 10,000 test pictures in the MNIST database. These are black-and-white pictures with 28 * 28 pixels, centered and normalized.

4.2   Model

In our work, we use the convolutional neural network model whose architecture is illustrated in Table 1.

Table 1. Architectures of our model.

DNN	Architecture
CNN	CNV (1,12,3), MaxPool (2), ReLU
	CNV (12,24,3), MaxPool (2), ReLU
	Linear (1176,64), ReLU, Linear (64,10)

4.3 Proposed Approach

The goal of our approach is to improve the robustness of deep neural networks against small perturbations. Deep learning classifiers are vulnerable to minor information perturbations. We try to understand the development of the decision boundary of deep neural networks, that is why we compute the distance of the images to the decision boundary and how this boundary evolves during the training of our classifier. The steps of our approach are as follows:

- Train our model on the MNIST database (Standard Training).
- During the training, we calculate the distance between the database images and the decision boundary, using the DeepFool technique.
- We calculate the distance between the database images and the decision boundary, using the DeepFool technique, but this time we differentiated between correctly and wrongly classified images (Standard Training).
- Train the model again, but this time, half of the images in each training batch are replaced by adversarial images generated using the DeepFool attack (Adversarial Training).
- During the Adversarial Training, we calculate the distance between the database images and the decision boundary, using the DeepFool technique.
- We calculate the distance between the database images and the decision boundary, using the DeepFool technique, but this time we differentiated between correctly and wrongly classified images. (Adversarial Training).
- Relevant observations to improve the robustness of deep neural networks.

5 Results and Discussion

All experiments were run on a Colab CPU with 12.69 GB of RAM and 107.72 GB of disk space. Due to its widespread use and simple built-in libraries, we chose Python 3 for programming. The Pytorch library is used to implement our approach. Our model was trained on 10 epochs with batch size $= 1$ and images of dimension 28 * 28.

The goal of our work is to track the l_2-norm border values d_2 for both training and validation during a deep convolutional neural network's training phase. Using the l_2-DeepFool technique, we can get these estimated image distances from the decision boundary. As a result,

$$d_2(x) = \|\text{DeepFool}_2(x)\|_2 \tag{6}$$

We will examine the average distance between the training and validation in each training period for every x image in order to deduce overall impressions. More specifically, we present for image data set D.

$$d_2^{avg} := \frac{1}{|D|} \sum_{x_i \in D} d_2(x_i) \tag{7}$$

The cardinality of the image group D is represented by $|D|$.

In numerous training phases, we will plot the average distance $d_2(x)$ for all images $x \in D$. These histogram representations provide a good sense of how the distance changes over the classifier's training. At MNIST, we do research. The study results for a deep neural network trained on the MNIST data set will be presented in the accompanying figures. Our results for the first 10 training epochs are presented in the graphs below. The MNIST model achieves good error rates following this limited training time. Finally, our findings point to three significant findings, which we will discuss in more detail in the sections that follow.

5.1 Standard Training

We understand that after training, at least a considerable part of the training and test images are around the decision boundary due to the neural networks' susceptibility to adversarial instances. To investigate this, we use MNIST to train a convolutional neural network (CNN), as illustrated in Fig. 1.

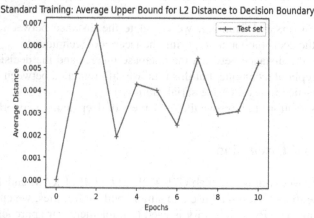

Fig. 1. CNN findings on the MNIST dataset, variation of the average distance d_2^{avg} over 1000 selected images from the test data.

It's worth noting that the distance values of the test and training sets are not significantly different. The summary is that the decision border comes near to the training and validation images. Regrettably, we are unable to provide a satisfactory and verifiable explanation for the decision boundary's behavior all through training at this time. As a result, the topic of why average distances are shrinking and how this evolution of the decision boundary can be stopped remains open.

5.2 Adversarial Training

On MNIST, we train CNN again, but this time in an aggressive manner. Like in the previous tests, the cross-entropy loss function and network designs were used. Half of the images in each training batch are replaced by adversarial images generated using the

DeepFool attack. Then, much like with test images until the decision border, we analyze the average distances of natural training again. The outcomes for CNN on MNIST are represented in Fig. 2. We see considerable differences from standard training outcomes, that is, the average distance d_2^{avg} remains at a pretty high value in comparison to classical training.

We've come to the conclusion that the gradual drop in average margins shown previously is not an unavoidable trend. In contrast to a traditionally trained model, the insertion of DeepFool adversarial instances into the training set results in a decision boundary that is farther away from natural data sources. Furthermore, the trials show that the durability of the adversialally trained classifier does not deteriorate completely with time. In general, we may say that DeepFool adversarial training produces models robust against small disturbances.

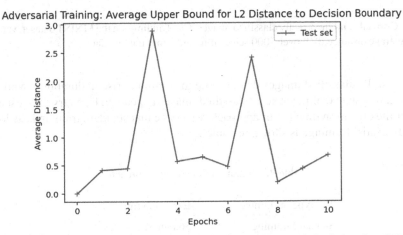

Fig. 2. Adversarial Training On the MNIST dataset, experimental findings of a CNN with DeepFool, variation of the average distance d_2^{avg} over 1000 randomly selected images.

5.3 Images that Have Been Classified Incorrectly

We trained CNN on the MNIST dataset as previously,but this time we differentiated between correctly and wrongly classified images. This leads to one simple conclusion: images that were incorrectly identified by CNN will typically be closer to the decision border than those that were successfully classified (see Fig. 3).

During standard model training, the proportion of wrongly classified images drifts closer to the decision boundary. This demonstrates how, as the training progresses, the decision boundary approaches natural images. As a result, images misclassified by a completely trained model will be very simple to move beyond the decision border with a minor disturbance.

Similarly, contradictory trained models exhibit the same decline in distance for wrongly classified images. An adversarial trained classifier's decision boundary tries to find a solution. It tries to attain good precision on the training set while also being robust in the vicinity of training images. The growing, or potentially steady, margins

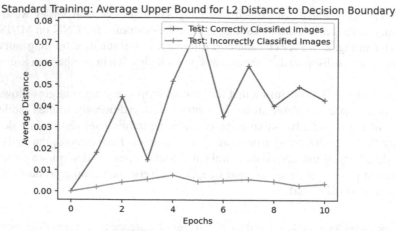

Fig. 3. Correctly and Incorrectly classified images: CNN findings on the MNIST dataset, variation of the average distance d_2^{avg} over 1000 selected images from the test data.

of successfully classified images demonstrate the desirable rise in durability, whilst the reduction in distance of incorrectly classified images appears to be an try to boost accuracy for the execution time in our approach, we notice that standard training lasts longer than adversarial training, as shown in Table 2.

Table 2. Execution time of our approach.

The type of training	Execution time
Standard training	36 min 31 s
Adversial training	15 min 18 s

6 Conclusion

In this experiment, the decision border of a deep learning model approaches the images in the training standard. The decision boundary continues to change even in the late phase of training, leading to the conclusion that completely trained models are sensitive to adversary disturbances, Adversarial training may be able to stop this undesirable decline in the orientation of margins to the decision border. Furthermore, there is a considerable difference in the distances to the decision border of successfully and wrongly identified images for trained classifiers. Images that have been wrongly classified are significantly nearer the decision border than images that have been successfully classified, and the boundary is approaching these images with time. This remark stays correct for both adversarial and standard training. Even while our findings help to improve comprehension of the decision boundary, there are so many unsolved research concerns.

Our findings give some insight on the issue of deep neural networks' sensitivity to adverse instances. We discovered, therefore, that this deficit in neural networks is caused by training rather than being specified by their initialization or construction. Then, from us, it's critical to investigate the impact of loss functions on distance. As a result, it's important to see if our findings hold true for increasingly complicated network designs and machine vision problems. Verifying the correctness of DeepFool's distance estimations is also critical. In this situation, we may obtain lower limits for input margins using additional powerful adversary strategies. This study can assist us in identifying verifiable explanations for previous observations. In summary, we believe that developing the decision boundary throughout training provides another dimension to a classifier's durability evaluation and gives us a better chance of detecting alternative causes for insufficient adversial resilience.

References

1. Yu, T., Hu, S., Guo, C., Chao, W.-L., Weinberger, K.Q.: A new defense against adversarial images: turning a weakness into a strength. In: NeurIPS (2019)
2. Tramer, F., Kurakin, A., Papernot, N., Goodfellow, I., Boneh, D., McDaniel, P.: Ensemble adversarial training: attacks and defenses. In: ICLR (2018)
3. Schott, L., Rauber, J., Bethge, M., Brendel, W.: Towards the first adversarially robust neural network model on MNIST. In: ICLR (2019)
4. Papernot, N., McDaniel, P., Wu, X., Jha, S., Swami, A.: Distillation as a defense to adversarial perturbations against deep neural networks. In: IEEE Symposium on Security and Privacy (2016)
5. Nar, K., Ocal, O., Sastry, S.S., Ramchandran, K.: Cross-entropy loss and low-rank features have responsibility for adversarial examples. arXiv:1901.08360 (2019)
6. Metzen, J.H., Kumar, M., Brox, T., Fischer, V.: Universal adversarial perturbations against semantic image segmentation. In: IEEE International Conference on Computer Vision (ICCV), pp. 2774–2783 (2017)
7. Madry, A., Makelov, A., Schmidt, L., Tsipras, D., Vladu, A.: Towards deep learning models resistant to adversarial attacks. In: ICLR (2018)
8. Kang, D., Sun, Y., Hendrycks, D., Brown, T., Steinhardt, J.: Testing robustness against unforeseen adversaries. arXiv:1908.08016 (2019)
9. Jiang, Y., Krishnan, D., Mobahi, H., Bengio, S.: Predicting the generalization gap in deep networks with margin distributions. In: ICLR (2019)
10. He, W., Li, B., Song, D.: Decision boundary analysis of adversarial examples. In: ICLR (2018)
11. Gu, S., Rigazio, L.: Towards deep neural network architectures robust to adversarial examples. In: ICLR (Workshop) (2015)
12. Ford, N., Gilmer, J., Carlini, N., Cubuk, D.: Adversarial examples are a natural consequence of test error in noise. In: ICML (2019)
13. Fawzi, A., Fawzi, H., Fawzi, O.: Adversarial vulnerability for any classifier. In: NeurIPS (2018)
14. Dodge, S., Karam, L.: A study and comparison of human and deep learning recognition performance under visual distortions. In: 2017 26th International Conference on Computer Communication and Networks (ICCCN), pp. 1–7. IEEE (2017)
15. Carlini, N., Wagner, D.: Towards evaluating the robustness of neural networks. In: IEEE Symposium on Security and Privacy, pp. 39–57 (2017)
16. Mickisch, D., Assion, F., Greßner, F., Günther, W., Motta, M.: Understanding the decision boundary of deep neural networks: an empirical study (2020)
17. Athalye, A., Carlini, N., Wagner, D.: Obfuscated gradients give a false sense of security: circumventing defenses to adversarial examples. In: ICML (2018)

18. Akhtar, N., Mian, A.: Threat of adversarial attacks on deep learning in computer vision: a survey. IEEE Access **6**, 14410–14430 (2018)
19. Volpi, R., Murino, V.: Addressing model vulnerability to distributional shifts over image transformation sets. In: ICCV (2019)
20. Szegedy, C., et al.: Intriguing properties of neural networks. In: ICLR (2014)
21. Shafahi, A., Huang, W.R., Studer, C., Feizi, S., Goldstein, T.: Are adversarial examples inevitable? In: ICLR (2019)
22. Schmidt, L., Santurkar, S., Tsipras, D., Talwar, K., Mdry, A.: Adversarially robust generalization requires more data. In: NeurIPS (2018)
23. Papernot, N., McDaniel, P.: Deep k-nearest neighbors: towards confident, interpretable and robust deep learning. arXiv:1803.04765 (2018)
24. Moosavi-Dezfooli, S.-M., Fawzi, A., Frossard, P.: Deepfool: a simple and accurate method to fool deep neural networks. In: CVPR (2016)
25. Metzen, J.H., Genewein, T., Fischer, V., Bischoff, B.: On detecting adversarial perturbations. In: Proceedings of 5th International Conference on Learning Representations (ICLR) (2017)
26. Johnson, J.: Deep, skinny neural networks are not universal approximators. In: ICLR (Poster) (2019)
27. Hendrycks, D., Dietterich, T.: Benchmarking neural network robustness to common corruptions and perturbations. In: ICLR (2019)
28. Guo, C., Rana, M., Cisse, M., van der Maaten, L.: Countering adversarial images using input transformations. In: ICLR (2018)
29. Goodfellow, I., Shlens, J., Szegedy, C.: Explaining and harnessing adversarial examples. In: ICLR (Poster) (2014)
30. Fawzi, A., Moosavi-Dezfooli, S.-M., Frossard, P., Soatto, S.: Classification regions of deep neural networks. arXiv:1705.09552 (2017)
31. Eykholt, K., et al.: Physical adversarial examples for object detectors. In: WOOT 2018 Proceedings of the 12th USENIX Conference on Offensive Technologies (2018)
32. Cisse, M., Bojanowski, P., Grave, E., Dauphin, Y., Usunier, N.: Parseval networks: Improving robustness to adversarial examples. In: ICML (2017)
33. Biggio, B., et al.: Evasion attacks against machine learning at test time. In: Blockeel, H., Kersting, K., Nijssen, S., Železný, F. (eds.) ECML PKDD 2013. LNCS (LNAI), vol. 8190, pp. 387–402. Springer, Heidelberg (2013). https://doi.org/10.1007/978-3-642-40994-3_25
34. Assion, F., et al.: The attack generator: a systematic approach towards constructing adversarial attacks. In: CVPR SAIAD (2019)

Facial Expression Recognition Using a Hybrid ViT-CNN Aggregator

Rachid Bousaid[1]([✉])[ID], Mohamed El Hajji[1,2][ID],
and Youssef Es-Saady[1][ID]

[1] IRF-SIC Laboratory, Ibn Zohr University, B.P 8106, Agadir, Morocco
{r.bousaid,y.essaady}@uiz.ac.ma
[2] CRMEF-SM, Avenue My Abdallah BP N° 106, Inezgane, Morocco
m.elhajji@crmefsm.ac.ma

Abstract. Facial Emotion Recognition (FER) is an important and challenging task in computer vision due to different issues such as quality of images, the correlation between same expression, computational complexity, and it requires a large amount of data. This paper presents a novel approach to the FER task. We are motivated by the success of Vision Transformer (ViT) and the Convolutional Neural Network (CNN) on image classification in general and facial emotion recognition.

The Swin Transformer (ST) is a hierarchical transformer that uses shifted windows to compute representation. The advantages of ST include limiting self-attention computing, and has linear computational complexity to image size. This paper studies and compares both ST and Deep CNN architecture when merged by different merging layers. The proposed approach is tested on the FER2013 and CK+ data sets. Experimental results demonstrate the high performance of the Average Merging Layer (AML), and our method outperforms state-of-the-art methods on FER2013 and CK+.

Keywords: Convolutional Neural Network (CNN) · Transformers · Facial Emotion Recognition (FER)

1 Introduction

Due to the high demand of FER systems in several application areas such as computer vision, FER has recently been a rich field for researchers in the field of artificial intelligence in general and machine learning in particular. P. Ekman and W. V. Friesen [1] distinguish seven basic facial expression classes: Angry, Disgust, Fear, Happy, Sad, Surprise, and Neutral. In their study, they find out that facial expressions of emotion are universal. Most of previous works in this field were based on these seven basic classes. In the literature, different approaches and methods are used for FER with face images such as SVM, Bayesian Network, and methods based on a descriptor like SIFT and its variants. Recently, the CNNs [2,3] have proven their efficiency in classifying and recognizing images, and they

© Springer Nature Switzerland AG 2022
M. Fakir et al. (Eds.): CBI 2022, LNBIP 449, pp. 61–70, 2022.
https://doi.org/10.1007/978-3-031-06458-6_5

have extended to the problematicality of FER. However, training a CNN requires a large amount of data and a high computational and memory cost. Moreover, the variation of lighting, pose skin color, and occlusion has a negative impact on the training of a deep CNN architecture. However, recognizing facial expressions with variant head poses, occlusion, etc. remains a challenging task for CNN. Over the past few years, ViT has shown success against occlusion and is suitable to deal with wild facial expression images [4]. As we are motivated by the success of transformer initially in NLP field then in computer vision, we propose a hybrid approach by combining Swin Transformer [5] and shallow DeepCNN. Several merging layers are investigated and compared. Figure 4 displays an overview of the proposed approach.

The rationale of this paper is to investigate the impact of combining ViT with CNN to increase the performance of FER, and compare different merging layers to choose the optimal merging that achieves a state of the art results on both FER-2013 and CK+ datasets.

2   Related Works

This section presents some relevant works on FER. Many works have applied CNN in the field of emotion recognition, and we are mainly concerned with two streams of related works: CNN-FER and Vit-FER.

In [6], Jason C. Hung et al. propose a strategy to test the effectiveness of transfer learning in a CNN model. Their method evaluates transfer learning using a basic emotion recognition model on FER2013 after pre-treatment and data augmentation. The results of the study demonstrate that the transfer-learning-based model shows good performance. [7] adjusts the hyper-parameters of their CNN architecture by following two approaches: keeping the number of layers and reducing the size of the filters or using a small size of the filters (3×3) in all layers by increasing the number of filters. Inspired by VGG16, the proposed model, encompasses 5 convolution layers, 5 of max-pooling layers, and 2 FC layers. The activation function is ReLu for convolution layers and softmax for classification. [8] combines the SIFT descriptor with a CNN architecture during the training and testing phases, compared to the other related works. In their work, the authors compare three models: CNN architecture, combination of SIFT and CNN, and combination of Dense SIFT and CNN. Accordingly, they build a final model by aggregating the outputs of the three models, using the average sum of probabilities. The combination of SIFT and CNN consists of transmitting the vector output from SIFT in a fully connected layer and then establishing a Dropout of 0.5 to have a vector that will be concatenated with the output of CNN in an FC layer. The models are tested on both FER2013 and CK+ databases. [9] offer a multi-view CNN for facial expression recognition with two databases, namely FER2013 and RAF. The model contains 8 layers of convolution, 4 layers of max pooling, and finally 2 FC layers. The activation function used is ReLu except for the last layer where softmax is used for the classification. In order to reduce the overfitting, they use a Dropout of 25%. This method combines two models, one of which uses the original images of FER2013 solely, whereas the other uses data augmentation. This combination takes the average of the output of the two networks as input for the softmax function of the last classification layer. The results achieved are 72.27% on FER2013 and 83.08% on RAF.

In [10] Mouath Aouayeb et al. combine a Vision Transformer with a Squeeze and Excitation (SE) block for FER task. SE is an attention mechanism but contains widely fewer parameters compared to self-attention block. This approach aims to extract local attention features by ViT and global relation from the extracted features by SE. In [4] Hanting Li et al. propose a transformer-based Mask and Vision Transformer (MVT). The basic idea of this approach is to generate a mask that can filter backgrounds and occlusions, then use a dynamic relabeling module. It therefore aims to reduce the impact of variant backgrounds, and the subjectiveness of annotators.

The main Challenges faced when using CNN for FER are computational complexity, image resolution, illumination variations, high intra-class variations, and high inter-class similarities introduced by facial appearance changes. To solve these issues, some works have thought about combining deep learning techniques into a hybrid system.

3   Method

The proposed solution consists of merging two models, namely Swin Transformer, and a DeepCNN that we are building from scratch. This section presents the proposed solution.

3.1   Overview of Swin Transformer

ST architecture [5] is depicted in Fig. 1. This architecture is mainly composed of a splitting module that splits an input image into patches, then four steps that produce a hierarchical representation:

- Step 1: The transformer blocks and linear embedding;

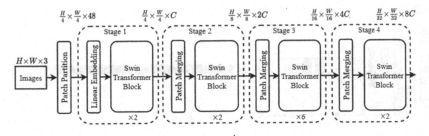

Fig. 1. The architecture of a Swin Transformer (ST) [5].

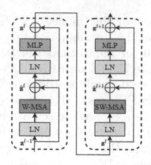

Fig. 2. Two successive ST Blocks [5].

- Step 2: Block of patch merging and feature transformation;
- Step 3 and step 4: The repetition of the step1 and step2 twice.

In the ST block, the authors replace standard Multi-head Self Attention (MSA) by shifted windows-based MSA module and applying a residual connection. Figure 2 displays two successive ST Blocks, LN means Layer Norm, W-MSA means MSA with standard windowing configuration, SW-MSA means MSA with shifted windowing configuration, and MLP means multi-layer perceptron.

The computational complexity of MSA and W-MSA are:

$$\Omega(MSA) = 4C^2(hw) + 2C(hw)^2 \tag{1}$$

$$\Omega(W - MSA) = 4C^2(hw) + 2CM^2(hw) \tag{2}$$

where hw is patch number, we note that the complexity is quadratic to hw in the first case (1) and linear in the second (2) when M and C are fixed.

Thanks to hierarchical representation, one of the advanced techniques for prediction, and linear computational complexity relative to image size, a Swin transformer can be adapted for different vision tasks, compared to ViTs which use single resolution feature maps.

Swin Transformer's performance for various vision problems motivated us to use it for the FER problem.

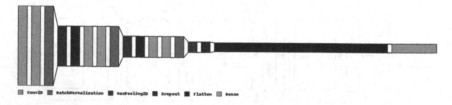

Fig. 3. Visualization of our DeepCNN architecture

3.2   Deep CNN Architecture

We designed our DeepCNN architecture from scratch as in [8]. It is shallow architecture with convolution layers and pooling layers.The number of filters is 32 for the first two convolution layers then the number is doubled every two convolution layers. We used kernel size $(3, 3)$ for convolution layers and $(2, 2)$ for pooling layers concerning the activation function we used Relu function defined as:

$$f(x) = max(0, x) \tag{3}$$

Figure 3 shows a visualization of our DeepCNN architecture. Dropout rate is $0.1, 0.1$ and 0.5 respectively. The output size is 1024.

3.3   Hybrid ViT-DeepCNN Model

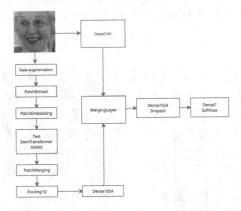

Fig. 4. Proposed architecture

There are several ways to merge models. In the current paper, we compare various merging layers to build an optimal model that merges Deep CNN and Transformer described above. The architecture we are testing is shown in Fig. 4.

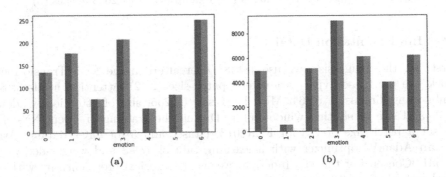

Fig. 5. Distribution of expressions on CK+ (a), Distribution of expressions on FER2013 (b)

4 Experiments

4.1 Data Sets

This approach evaluated on the Extended Cohn-Kanade (CK+) [11] and FER 2013 datasets that are widely used in FER works.

Extended Cohen Kanade dataset (CK+) is a controlled dataset. The version we opted for contains 981 images distributed as 73% for training, 18% for validation, and 8% for testing. It divides into seven basic expressions which are: Disgust, Fear, Angry, Happy, Sad, Surprise, and Neutral. Figure 5a shows the distributions of the expressions, and Fig. 6a shows examples of CK+ images.

FER 2013 dataset is a wild facial expression image dataset, and the version we opted for contains 35187 images labeled with one of the seven basic expressions, and includes 80% for training, 10% for validation, and 10% for testing. Figure 5 a shows the distributions of the expressions, and Fig. 6b shows examples of FER 2013 images.

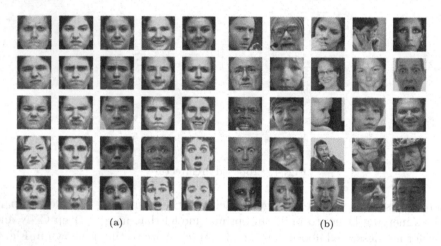

(a)   (b)

Fig. 6. Examples of CK+ images (a), Examples of FER2013 images (b)

4.2 Implementation Details

First, for the image size, we use 48*48. Thereafter, in the Swin Transformer block, we use these hyper-parameters: patch-size (2, 2), attention heads of 8, Embedding dimension of 64, MLP layer size of 256, size of attention window of 2, and size of shifting window of 1. During the training of DeepCNN-ST, we set a batch-size of 128 for FER2013 dataset and 5 for CK+ dataset. We use an AdamW optimizer with a learning rate of $1e-3$ and weight-decay of 0.0001. Concerning the loss function, we use categorical cross entropy with a label smoothing of 0.1.

To ensure a fair comparison and evaluation, the experiments are conducted with the same model for the two datasets. All the models are trained with 10-fold-cross-validation. We use all the images in the dataset. The model is developed using Keras and Tensorflow framework on Kaggle cloud software.

4.3   Experimental Results for Data Sets

This section presents the results of several types for model merging that we have applied to combine DeepCNN and Swin Transformer. The accuracies of all the methods are displayed in Table 1, and Table 2.

The Effect of Merging Type

In order to compare the performance using different merging types. The proposed method computes the average scores for all folds, and is tested on two datasets; i.e. CK+ and FER2013. As shown in Tables 1 and 2, we found out that using the Average merging layer performs better on the two datasets.

Figure 7 displays the normalized confusion matrixes for the test data using the proposed model and FER2013 dataset. Figure 7a in case of using Average merging, and Fig. 7b in case of using concatenation merging, the two models that have the best average scores. It demonstrates that class 3 (Happy) has a good prediction result. One possible reason is the distribution of labeling images in the dataset. As shown in Fig. 5b, class 3 has a high amount of samples.

Table 1. Performance of the models on CK+

Architecture	Best-accuracy	Average scores for all folds	Epochs
Swin Transformer	96.93%	93.49%	20 * 10
Dcnn & ST by Maximum-ML	100%	98.88%	20 * 10
Dcnn & ST by Concatenate-ML	100%	98.28%	20 * 10
Dcnn & ST by Average-ML	**100%**	**99.89%**	20 * 10
Dcnn & ST by Dot-ML	98.97%	77.18%	20 * 10
Dcnn & ST by Multiply-ML	100%	99.59%	20 * 10

Table 2. Performance of the models on FER2013

Architecture	Best-accuracy	Average scores for all folds	Epochs
Swin Transformer	50.75%	47, 38%	20 * 10
Dcnn & ST by Maximum-ML	90.30%	81.96%	20 * 10
Dcnn & ST by Concatenate-ML	89.93%	82.28%	20 * 10
Dcnn & ST by Average-ML	**90.55%**	**83.88%**	20 * 10
Dcnn & ST by Dot-ML	60.06%	55.17%	20 * 10
Dcnn & ST by Multiply-ML	84.83%	79.65%	20 * 10

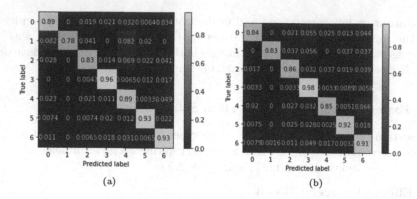

Fig. 7. Confusion matrix for fer2013: with Average-Merging (a), with Concat-Merging (b)

4.4 Comparison with State-of-the Arts

This part uses the best accuracy in 10-cross-validation to compare the proposed approach with the competing methods. As shown in Tables 3 and 4, the proposed method surpasses state-of-the-art in the two datasets. For a comprehensive comparison, we compute average scores for all folds to obtain the mean accuracy for each dataset and merging layer choice. The Average Merging layer achieves the best mean accuracies, 83.88% for FER2013 and 99.89% for CK+.

Table 3. Comparison with state-of-the arts on CK+

Method	Year	Accuracy
CNN-SIFT aggregator [8]	2017	99.1%
eXnet [12]	2020	96.75%
ViT+SE [10]	2021	99.80%
ST+DeepCNN (our)	2022	**100%**

Table 4. Comparison with state-of-the arts on FER2013

Method	Year	Accuracy
CNN-SIFT aggregator [8]	2017	73,4%
CNN [13]	2020	65,77%
Dense-FaceLiveNet [6]	2019	70,02%
CNN+LS-ReLu [14]	2020	90,16%
MVT [4]	2021	89.22%
FER-VT [15]	2021	90.04%
ST+DeepCNN (our)	2022	**90.55%**

5   Conclusion

This study first analyzes the hybrid Swin Transformer with DeepCNN and their performance effect on the FER task. Then, we compare the merging layer types to choose the optimal one. Experimental results demonstrate the high performance of the Average merging layer. Finally, it is concluded that the transformer and convolutional neural network could complement each other in improving the accuracy results on both FER-2013 and CK+ datasets. We therefore aim to extend the merging models of ViT and CNN architecture to propose a solution for the real-time micro-expressions recognition issue potential future works.

Acknowledgments. This work was supported by the Ministry of Higher Education, Scientific Research and Innovation, the Digital Development Agency (DDA), and the CNRST of Morocco (project 22).

References

1. Ekman, P., Friesen, W.V.: Constants across cultures in the face and emotion. J. Pers. Soc. Psychol. **17**(2), 124–129 (1971). https://doi.org/10.1037/h0030377
2. LeCun, Y., Bottou, L., Bengio, Y., Haffner, P.: Gradient-based learning applied to document recognition. Proc. IEEE **86**(11), 2278–2323 (1998). https://doi.org/10.1109/5.726791
3. Krizhevsky, A., Sutskever, I., Hinton, G.E.: ImageNet classification with deep convolutional neural networks. Commun. ACM **60**(6), 84–90 (2017). https://doi.org/10.1145/3065386
4. Li, H., Sui, M., Zhao, F., Zha, Z., Wu, F.: MVT: Mask Vision Transformer for Facial Expression Recognition in the Wild (2021). arXiv:2106.04520
5. Liu, Z., et al.: Swin Transformer: Hierarchical Vision Transformer using Shifted Windows (2021). arXiv:2103.14030
6. Hung, J.C., Lin, K.C., Lai, N.X.: Recognizing learning emotion based on convolutional neural networks and transfer learning. Appl. Soft Comput. J. **84**, 105724 (2019). https://doi.org/10.1016/j.asoc.2019.105724
7. Rzayeva, Z., Alasgarov, E.: Facial emotion recognition using deep convolutional neural networks. Int. J. Adv. Sci. Technol. **29**(6 Special Issue), 2020–2025 (2020)
8. Connie, T., Al-Shabi, M., Cheah, W.P., Goh, M.: Facial expression recognition using a hybrid CNN–SIFT aggregator. In: Phon-Amnuaisuk, S., Ang, S.-P., Lee, S.-Y. (eds.) MIWAI 2017. LNCS (LNAI), vol. 10607, pp. 139–149. Springer, Cham (2017). https://doi.org/10.1007/978-3-319-69456-6_12
9. Alfakih, A., Yang, S., Hu, T.: Distributed computing and artificial intelligence. In: 16th International Conference, Multi-view Cooperative Deep Convolutional Network for Facial Recognition with Small Samples Learning, vol. 290 (2019). https://doi.org/10.1007/978-3-030-23887-2
10. Aouayeb, M., Hamidouche, W., Soladie, C., Kpalma, K., Seguier, R.: Learning Vision Transformer with Squeeze and Excitation for Facial Expression Recognition, pp. 1–13 (2021). arXiv:2107.03107
11. Lucey, P., Cohn, J.F., Kanade, T., Saragih, J., Ambadar, Z., Matthews, I.: The extended cohn-kanade dataset (ck+): a complete dataset for action unit and emotion-specified expression. In: 2010 IEEE Computer Society Conference on Computer Vision and Pattern Recognition - Workshops, pp. 94–101 (2010). https://doi.org/10.1109/CVPRW.2010.5543262

12. Riaz, M.N., Shen, Y., Sohail, M., Guo, M.: eXnet: an efficient approach for emotion recognition in the wild. Sensors (Switzerland) **20**(4), 1087 (2020). https://doi.org/10.3390/s20041087

13. Agrawal, A., Mittal, N.: Using CNN for facial expression recognition: a study of the effects of kernel size and number of filters on accuracy. Visual Comput. **36**(2), 405–412 (2019). https://doi.org/10.1007/s00371-019-01630-9

14. Wang, Y., Li, Y., Song, Y., Rong, X.: The influence of the activation function in a convolution neural network model of facial expression recognition. Appl. Sci. **10**(5), 1897 (2020). https://doi.org/10.3390/app10051897

15. Huang, Q., Huang, C., Wang, X., Jiang, F.: Facial expression recognition with grid-wise attention and visual transformer. Inf. Sci. (Ny). **580**, 35–54 (2021). https://doi.org/10.1016/j.ins.2021.08.043

Machine Learning Approach to Automate Decision Support on Information System Attacks

Younes Wadiai and Mohamed Baslam[(✉)] [iD]

TIAD Laboratory, Sciences and Technology Faculty, Sultan Moulay Slimane University, Beni Mellal, Morocco
{younes.wadiai,m.baslam}@usms.ma

Abstract. As more software solutions are now cloud based taking advantage of the powerful computing performance of remote servers and super computers, the machine learning industry is also switching to this technology providing promising solutions such as Google Cloud Artificial Intelligence, Amazon Web Services, and Microsoft Azure Machine Learning. With the adoption of the cloud technology for nowadays computer transactions and operations, a cloud IDS solution that can compete with the emerging technology challenges is crucially needed to help network administrators secure data and prevent any intrusions. The machine learning approaches often require high computing performance and gigantic memory space to process mega datasets and come up with better prediction results. This paper introduces a new aspect of using cloud based machine learning solution as an online computing resource for the application of machine learning concepts to predict intrusions in IDS systems based on network packet behavior, while the traditional way is to use local computer resources through data mining solutions such as Weka or Orange. We used Microsoft Azure Machine Learning Studio along with CSE-CIC-IDS 2018 dataset from the Canadian Institute for Cyber Security to apply various techniques and algorithms to come up with a powerful network model. The aim of this paper is to explain how a cloud-based data mining tool can be used for its better performance and high accuracy for data mining and building a strong intrusion detection system. As a start point, we used a saved dataset that contains a collection of anomaly detection records applied on an IDS while various attributes are registered. To differentiate between normal and anomalous traffic, two profiles are used: B-profile and M- profile to generate benign and malicious traffic respectively in the network.

Keywords: Intrusion detection system · Machine learning · Deep learning · Neural network · Azure ML Studio · Multi Class Regression · Multi Class Neural Network

1 Introduction

As artificial intelligence is taking more space in nowadays computing world, and with the emergence of new web attacks that use untraditional techniques and can penetrate

M. Fakir et al. (Eds.): CBI 2022, LNBIP 449, pp. 71–81, 2022.
https://doi.org/10.1007/978-3-031-06458-6_6

unnoticedly through existing security solutions, the usage of machine learning to help securing the exponential growth of networks and data become crucial. In this paper, we are suggesting a new way of enhancing the performance of IDS detection rate through building a new prediction model using cloud-based machine learning tool: Microsoft Azure Machine Learning Studio. We used CSE-CIC-IDS-2018 dataset which is a combination of various attack scenarios on a production network. Once the dataset is downloaded and combined in one master dataset, the exploitation comes next step using the normalization and cleaning tools provided in azure ML studio. After the summarization, filtering, and cleaning the dataset, the experiment was performed using various supervised and unsupervised algorithms including: Multi Class Logistic Regression [9], and Multi Class Neural Network [10]. The found results showed the outstanding of deep learning algorithms in terms of accuracy and prediction. The structure of this paper is presented as follows: Section 2 is covering some related work where machine learning was applied on the IDS systems and the CSE-CIC-IDS-2018 dataset describing the strengths and weaknesses of each study, while Sect. 3 will introduce our approach with detailed explication of the preparation and application phase on the cloud platform. Section 4 is dedicated to discuss findings and analyze results of each applied method, and lastly Sect. 5 concludes the work and recommends directions for future research.

2   Related Work

Since the intrusion detection system can be identified as any combination of software/hardware mechanism capable to detect malicious traffic in a network to prevent damage, malfunction, or compromise it, this section of the paper introduces previous works on IDS systems using CSE-CIC-IDS-2018 dataset and its results, in [11] for example, the general outcome of the performed survey resulted that poor effort was made in most research to clean big data before using it to build a prediction model, the usage of an imbalanced class leads to inaccurate experiment tests and skews the results. Another study that focuses on intrusion detection using multi-architectural module can be found in [10] where the paper introduced a technique that consist of a feed-forward module to decrease the high rate of the false positive alarms. Another study was on the intrusion detection using deep learning algorithms is [11] where various frameworks were used like Keras, TensorFlow, PyTorch, and fast.ai where the last framework provided 99% accuracy rate, however the usage of one classifier is a drawback of this study. As a try to generalize a trained model and apply it on a different dataset/platform, the [12] paper proposed the usage of twelve various supervised machine learning algorithms from different families and apply them on two experiment datasets. In order to tackle the imbalance in the class of 2018 dataset, the [13] paper explains how the LightGBM classifier was used to address high values of attributes and instances of the dataset. The novel in [6] introduced the usage of an artificial neural network where two-layer MLP algorithms trained and applied on the 2018 dataset to get over 99% accuracy score with a false positive rate of 0.001. A great paper in [14] presents the application of cloud computing on the CSE-CIC-IDS-2018 and Python and Scikit-learn [15], the experiment showed the outstanding of the MLP classifier over the other classifiers: Random Forest [16], kNN [17], SVM [18], Adaboost [19], and Naive Bayes [20] with a perfect AUC [21] score of 1 and 99% accuracy.

3 Proposed Approach

In this section, we will present the work environment and how the dataset was prepared and uploaded to azure studio platform followed by other steps and techniques used to get the wanted results.

CSE-CIC-IDS-2018 Dataset in Azure ML Studio:
The CSE-CIC-IDS2018 dataset by the Canadian Institute for Cyber Security is a mega dataset widely used in various research articles for intrusion detection and prevention systems.

The dataset simulates different scenarios of attacks including: DoS Denial of Service, DDoS Distributed Denial of Service, Heartbleed bug, exhaustive search known as brute force, botnet attack carried by malwares, web attack, and an inside infiltration of the production network. The attacks were performed by fifty devices targeting a production network containing thirty servers and over four hundred computers, among five departments. By using the CICFlowMeter-V3, 80 attributes are recorded in the dataset.

Figure 1 shows the architecture of the network.

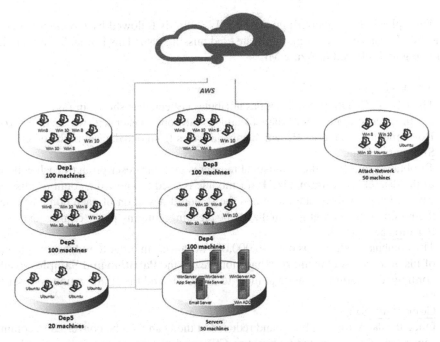

Fig. 1. Architecture of the network

After downloading the dataset from the amazon web service where it's stored as follows:

- Friday-02-03-2018_TrafficForML_CICFlowMeter
- Friday-16-02-2018_TrafficForML_CICFlowMeter
- Friday-23-02-2018_TrafficForML_CICFlowMeter
- Tuesday-20-02-2018_TrafficForML_CICFlowMeter
- Thursday-01-03-2018_TrafficForML_CICFlowMeter
- Thursday-15-02-2018_TrafficForML_CICFlowMeter
- Wednesday-14-02-2018_TrafficForML_CICFlowMeter
- Wednesday-21-02-2018_TrafficForML_CICFlowMeter
- Wednesday-28-02-2018_TrafficForML_CICFlowMeter
- Thursday-22-02-2018_TrafficForML_CICFlowMeter

The distributed dataset is then combined into one master dataset that contains all days of the attacks.

A. The Preprocessing Phase (Preparing Work Environment):

The upload of the dataset into Azure ML Studio is followed by the preprocessing phase to clear noisy, irrelevant, duplicate, and missing data. This is doable through the following tools offered in Azure ML:

1. Add Rows:
   The CSE-CIC-IDS2018 dataset has attributes collected as shown in Fig. 2.
   In order to combine all days of attacks into one master list, we used the **Add Rows** module in Azure ML which appends rows from one dataset to the other.
2. Summarize Data:
   To obtain a set of standard statistical measures with the description of each column in the dataset, the **Summarize Data** module is used to know the count of missing values, unique values, the mean and the standard deviation of each column. Once the report of the dataset is visualized, the cleaning, filtering phase comes next.
3. Partition and Sample:
   The combined dataset has over 16,000,000 instances, in order to reduce the number of instances to avoid memory exhaustion we use the **Partition and Sample** module configured to sampling mode at the rate of 0.0001 and the stratification split set to on.
4. Convert to CSV:
   Once the dataset is combined and reduced, it then can then be converted to comma separated values using the **Convert to CSV** module for exportation or to be shared with a Python script module later. Figure 3 Shows the combination and reduction phase.
5. Select Columns in Dataset:
   The Tuesday 20-02-2018 record has 84 instances instead of 80, we used the **Select Columns in Dataset** module to add its row to the master dataset. Flow ID, Src IP, Src Port, and Dst IP removed.

The CSE-CIC-IDS2018 Attributes

fl_dur	bw_iat_std	fw_pkt_blk_avg
tot_fw_pk	bw_iat_max	fw_blk_rate_avg
tot_bw_pk	bw_iat_min	bw_byt_blk_avg
tot_l_fw_pkt	fw_psh_flag	bw_pkt_blk_avg
fw_pkt_l_max	bw_hdr_len	bw_blk_rate_avg
fw_pkt_l_min	fw_pkt_s	subfl_fw_pk
fw_pkt_l_avg	bw_pkt_s	subfl_fw_byt
fw_pkt_l_std	pkt_len_min	subfl_bw_pkt
Bw_pkt_l_max	pkt_len_max	subfl_bw_byt
Bw_pkt_l_min	pkt_len_avg	fw_win_byt
Bw_pkt_l_avg	pkt_len_std	bw_win_byt
Bw_pkt_l_std	pkt_len_va	Fw_act_pkt
fl_byt_s	fin_cnt	fw_seg_min
fl_pkt_s	syn_cnt	atv_avg
fl_iat_avg	rst_cnt	atv_std
fl_iat_std	pst_cnt	atv_max
fl_iat_max	ack_cnt	atv_min
fl_iat_min	urg_cnt	idl_avg
fw_iat_tot	cwe_cnt	idl_std
fw_iat_avg	ece_cnt	idl_max
fw_iat_std	down_up_ratio	idl_min
fw_iat_max	pkt_size_avg	**Label**
fw_iat_min	fw_seg_avg	
bw_iat_tot	bw_seg_avg	
bw_iat_avg	fw_byt_blk_avg	

Fig. 2. The CSE-CIC-IDS2018 attributes

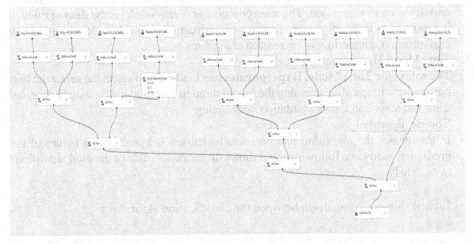

Fig. 3. Combination and reduction of dataset in Azure machine learnign studio.

6. Remove Duplicate Rows:
   As its name indicate, the **Remove Duplicate Rows** module eliminates the duplicated records from the dataset.

7. Clean Missing Data:
   To prevent problems that an arise while training the model, the **Clean Missing Data** module plays the role of removing rows and columns with missing values or replacing missing values with a placeholder or a mean value.

8. Normalize Data:
   The **Normalize Data** module is used to adjust numeric values in columns to use a common scale. The used method in our case is Zscore which convert all values to a Zscore.
       The Zscore is calculated as:

$$Z = x - \mu/\sigma$$

where Z represents the standard score, X the observed vale, μ mean of the sample. And σ is the standard deviation of the sample.

B. Data Exploitation (split dataset, train, score, and evaluate model):

1. Split dataset:
   The **Split Dataset** module divides the dataset in two sections, the first set is for training and the second is for testing, the percentage of the sets is 70% and 30% respectively.

2. Train Model:
   The **Train Model** is the phase to choose the algorithm needed to train the dataset classification or regression. The main component of this model is the **ilearner** which is produced as a specific binary format with instructions to perform learning through classification, clustering, or regression algorithms.

3. Tune Model Hyperparameters:
   The role of the **Tune Model Hyperparameters** module is to select the select the best parameter settings after scanning the model through two modes of tuning: Integrated train and tune, and Cross Validation with tuning.

4. Choose algorithm:
   In this phase, the algorithm that needs to be chosen is based on the nature of the prediction needs, the following table shows the characteristics of the used algorithms in our study.

Table 1 shows a comparison between the classification algorithms.

Table 1. Comparison of classification algorithms.

Algorithm	Accuracy	Training time	Linearity	Parameters
Multiclass regression	Good	Fast	Yes	4
Multiclass neural network	Good	Moderate	No	8

5. Score model:

   The **Score Model** is the phase of scoring the prediction performed in the previous training phase. The score can be in different format based on the predicted value, it can be a numeric value or a class object.

6. Evaluate model:

   The **Evaluate Model** is the metrics of the performance of the built model where results of the probable predictions are measured of a classification or a regression based on standard metrics.

Multiclass Logistic Regression:

Multiclass Logistic Regression in Azure ML Studio is a supervised machine learning algorithm that uses classification regression to predict multiple values. Since the logistic regression is a supervised machine learning method, it then requires a labeled dataset as an input to a train module where the t train module can then be used to predict values for new inputs. The Multiclass regression is an extension of the logistic regression which increases natural support to solve multiclassification issues. Figure 4 shows the schematic of the logistic regression classifier.

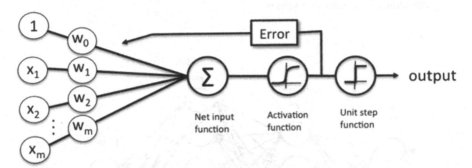

Fig. 4. Schematic of a logistic regression classifier

Softmax Regression Formula:

The Softmax regression formula also known as the multinomial logistic or the maximum entropy is used to compute the generalization of the logistic regression and it is expressed as:

$$P\left(y = j \middle| z^{(i)}\right) = \phi_{soft\,max}\left(z^{(i)}\right) = \frac{e^{z^{(i)}}}{\sum_{j=0}^{k} e^{z_k^{(i)}}}$$

where z is defined as the net input:

$$z = w_0 x_0 + w_1 x_1 + \ldots + w_m x_m = \sum_{l=0}^{m} w_1 x_1 = \mathbf{w}^T \mathbf{x}.$$

Multiclass Neural Network:
A neural network is a grid of interconnected layers, starting with layer number one where the input takes place, and finishing by the last layer where the output is. Between the input and output layers, there is a set of interrelated layers through nodes.

Figure 5 shows the different layers of a neural network where input, hidden, and output layers are shown.

The mathematical formula used in a neural network function is:

$$Z = \text{Bias} + W_1 X_1 + W_2 X_2 + \ldots + W_n X_n$$

where W represents the weights.
X is the input and it's an independent variable.
W_0 is the intercept which equals to the bias.
Z is the denotation of the graphical representation.

This Multi Classification algorithm is used because it's:

- Supervised learning method.
- Requires a tagged dataset that includes a label column.
- Label columns have multiple values.

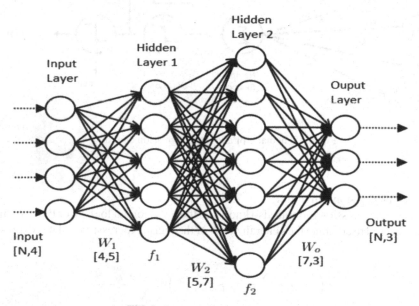

Fig. 5. Layers of a neural network

4 Results and Discussion

In order to perform an efficient comparison of the results, the comparison needs to be done between two similar models only, in our case, after building the evaluation model in Azure Cloud Studio as indicated in Fig. 6.

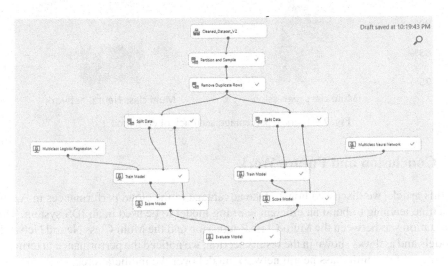

Fig. 6. Evaluating the accuracy of both models

The experiment is then based on comparing results of Multi Class Neural Network and Multi Class Regression and the following findings are noticed:

The overall accuracy of the Multiclass regression model reached 85%, with a micro averaged precision of 0.40.

On the other hand, the average accuracy of the multiclass neural network is 97% and a micro averaged precision scale of 0.70.

By analyzing other evaluation parameters, we also find that the Multi Class Neural Network has 10% fewer error values compared to the Multi Class Regression model.

It is obvious that the multiclass neural network model outstands the multiclass regression and provides better accuracy and prediction model. Figure 7 Shows the outperforming of the multiclass regression model over the multiclass neural network in terms of accuracy.

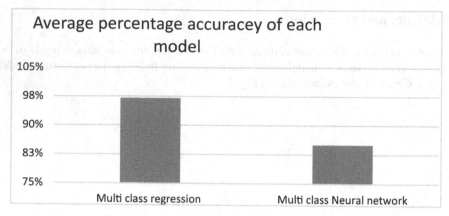

Fig. 7. Average percentage accuracy of each model

5   Conclusion and Future Work

In this article, we discussed how to take advantage of the cloud performances in Azure machine learning to build an efficient learning model to be used in an IDS system. The evaluation was between the Multi Class Regression and the Multi Class Neural Network models and as it was shown in the results section, we noticed the performance in terms of accuracy of the multiclass neural network model over the multiclass regression model, the neural network model gave us a result of 97% accuracy which allows us to accurately expect malicious traffic passing over an IDS. By exploring the Azure features, we were able to control the parameters of each model separately and measure its accuracy and error value, and again the Multi Neural Network model bypasses the Multi Regression one. The future work will include a new way to automatically feed the collected traffic from a production IDS in a network into a model and then orient the IDS about the action to be taken to protect the network in real time. This will be done thanks to the usage of web services feature found in Azure Studio Cloud.

References

1. Kumar, M.: An incorporation of artificial intelligence capabilities in cloud computing. Int. J. Eng. Comput. Sci. **5**(11), 19070–19073 (2016)
2. Cloud, Amazon Elastic Compute. Amazon web services (2011). Retrieved 9 Nov 2011
3. Mund, S.: Microsoft azure machine learning. Packt Publishing Ltd (2015)
4. Hall, M., et al.: The WEKA data mining software: an update. ACM SIGKDD Explor. Newsl. **11**(1), 10–18 (2009)
5. Demšar, J., et al.: Orange: data mining toolbox in Python. J. Mach. Learn. Res. **14**(1), 2349–2353 (2013)
6. Kanimozhi, V., Jacob, T.P.: Artificial intelligence based network intrusion detection with hyper-parameter optimization tuning on the realistic cyber dataset CSE-CIC-IDS2018 using cloud computing. In: 2019 International Conference on Communication and Signal Processing (ICCSP). IEEE (2019)

7. Panigrahi, R., Borah, S.: A statistical analysis of lazy classifiers using canadian institute of cybersecurity datasets. In: Borah, S., Emilia Balas, V., Polkowski, Z. (eds.) Advances in Data Science and Management. LNDECT, vol. 37, pp. 215–222. Springer, Singapore (2020). https://doi.org/10.1007/978-981-15-0978-0_21

8. Kanimozhi, V., Jacob, T.P.: Artificial intelligence outflanks all other machine learning classifiers in network intrusion detection system on the realistic cyber dataset CSE-CIC-IDS2018 using cloud computing. ICT Express 7(3), 366–370 (2021)

9. Gasso, G.: Logistic regression (2019)

10. Ou, G., Murphey, Y.L.: Multi-class pattern classification using neural networks. Pattern Recognit. 40(1), 4–18 (2007)

11. Leevy, J.L., Khoshgoftaar, T.M.: A survey and analysis of intrusion detection models based on CSE-CIC-IDS2018 big data. J. Big Data 7(1), 1–19 (2020). https://doi.org/10.1186/s40 537-020-00382-x

12. Sharafaldin, I., Lashkari, A.H., Ghorbani, A.A.: Toward generating a new intrusion detection dataset and intrusion traffic characterization. In: ICISSp, pp. 108–116 (2018)

13. Hua, Y.: An efficient traffic classification scheme using embedded feature selection and LightGBM. In: 2020 Information Communication Technologies Conference (ICTC). IEEE (2020)

14. Kanimozhi, V., Jacob, T.P.: Calibration of various optimized machine learning classifiers in network intrusion detection system on the realistic cyber dataset CSE-CIC-IDS2018 using cloud computing. Int. J. Eng. Appl. Sci. Technol. 4(6), 209–213 (2019)

15. Kramer, O.: Scikit-learn. In: Kramer, O. (ed.) Machine learning for evolution strategies, pp. 45–53. Springer, Cham (2016). https://doi.org/10.1007/978-3-319-33383-0_5

16. Pal, M.: Random forest classifier for remote sensing classification. Int. J. Remote Sens. 26(1), 217–222 (2005)

17. Yigit, H.: A weighting approach for KNN classifier. In: 2013 International Conference on Electronics, Computer and Computation (ICECCO). IEEE (2013)

18. Rueping, S.: SVM classifier estimation from group probabilities. In: ICML (2010)

19. Schapire, R.E.: Explaining adaboost. In: Schölkopf, B., Luo, Z., Vovk, V. (eds.) Empirical inference, pp. 37–52. Springer, Heidelberg (2013). https://doi.org/10.1007/978-3-642-411 36-6_5

20. Webb, G.I., Keogh, E., Miikkulainen, R.: Naïve Bayes. Encyclopedia Mach. Learn. 15, 713–714 (2010)

21. Huang, J., Ling, C.X.: Using AUC and accuracy in evaluating learning algorithms. IEEE Trans. Knowl. Data Eng. 17(3), 299–310 (2005)

Deep Reinforcement Learning for Bitcoin Trading

Bouchra El Akraoui[✉][iD] and Cherki Daoui[iD]

Laboratory of Information Processing and Decision Support,
Sultan Moulay Slimane University, B.P. 523, Beni Mellal, Morocco
b.elakraoui@usms.ac.ma

Abstract. Artificial intelligence (AI) is showing its success in various types of applications. Motivated by this trend, automatic trading has taken a keen interest in applying of artificial intelligence methods to predict the future price of a financial asset to overcome trading challenges including asset price fluctuations and dynamics, Investors must therefore understand when it is appropriate to use the optimal strategy that maximizes their investment return. But achieving a perfect strategy is difficult for an asset with a complex and dynamic price. To overcome these challenges, In this study, we apply a new rule-based strategy technique to train one of the successful machine learning algorithms, known as Deep Reinforcement Learning (DRL) for bitcoin trading. Our proposed method is based on dueling double deep q-learning networks, proximal policy optimization, and advantage actor-critic to achieve an optimal policy. The profit reward functions and Sharpe ratio are used to assess the proposed DRL. The results of the experiments demonstrate that combining three agents is the most efficient strategy for automatic bitcoin trading.

Keywords: Markov Decision Process · Deep Reinforcement Learning · Dueling double deep q-learning networks · Proximal policy optimization · Advantage actor-critic

1 Introduction

Researchers have been paying close attention to cryptocurrencies in recent years as a result of the significant rise in Bitcoin prices, which has led to a series of opportunities for traders to speculate on market prices, expanding research on financial trading, and suggesting a variety of methods for trading in the markets. One of the most difficult aspects of automatic trading is predicting the best strategies to maximize rewards. it is difficult to devise a profitable strategy within a dynamic and complicated price of a financial asset. Especially since bitcoin price fluctuations are erratic and are more volatile than other major asset classes, which makes them difficult to predict. Therefore, the investors need to understand the appropriate time to sell or buy the considered asset or hold the current position to maximize the return on their investment. Price prediction

© Springer Nature Switzerland AG 2022
M. Fakir et al. (Eds.): CBI 2022, LNBIP 449, pp. 82–93, 2022.
https://doi.org/10.1007/978-3-031-06458-6_7

theory is the decision-making process that is one of the main topics of discussion in finance that aims to maximize return while minimizing risk. One of the classical methods is carried out in two stages as mentioned in [22]. The stock prices' covariance matrix and predicted returns are computed first. The optimal asset allocation is therefore determined by either maximizing profit for a given investment risk or avoiding risks for a variety of returns. Following the optimal asset allocation, the profitable trading strategy is subsequently discovered. This method may be challenging to apply if the manager decides to change the decisions made at each time step and considers the transaction cost. For this purpose, the agent requires more than a price prediction model [11]. Several approaches have been proposed to tackle the issue of automated trading, especially, when the data had become available and market information had become more accessible [8]. The most efficient procedures that are based on optimization require deeper knowledge [12]. Another method to address the challenge of crypto-currency trading based on Reinforcement Learning (RL), is considered one of the effective methods of the Markov Decision Process (MDP) for finding the best strategy to solve these problems. However, it is now a pretty common tool for decision-making planning and learning, as well as determining optimal strategies in a partially known or unknown system. An MDP is a frequent means of simulating such a situation. The current state in the financial environment could be the actual price, current return, as well as some forecasting index values, and the next state he or she will end up in after taking the action are all provided in an MDP. The goal of RL from a financial standpoint is to determine the optimal strategy for a given situation that maximizes the expected discounted reward over a set of states. Though due to the "curse of dimensionality" of large state spaces, the RL model's scalability is constraine [1,9,27]. With the advent and tremendous progress of Deep Learning (DL), we have been able to create a variety of algorithms capable of fully automating complex tasks such as the recognition of images and classification [21]. Motivated by the mentioned challenges, we present in this paper, a recently successful and outstanding combination of the above methods (RL and DL frameworks), to generate automatic trading systems using a new research route named Deep Reinforcement Learning, which recommends taking the right action and achieves tremendous performance in trading by outperforming the traditional methods to maximize the rewards and get the best result in the complex, dynamic and high dimensional financial environment. This innovative method is based on the famous Deep Q-Network (DQN) framework, which is adapted to the specific challenge of automatic trading. In trading financial models, the proposed technique faces the following challenges:

i) Dynamic systems: the proposed models must be able to adjust to market changes fast. Even though financial markets are so volatile, and to cope with non-linear models of financial markets;

ii) The models must be flexible to every asset in diverse markets: Using the history of asset-specific price data, the suggested models are capable of learning suitable trading strategies for asset classes in diverse economies.

iii) One of the key elements of the suggested approaches for learning a good trading strategy for a certain financial asset is the feature extraction step. The performance of the learned trading rules is directly influenced by the quality of the extracted characteristics.

iv) Risk management: It is a key element of the financial market, which is afflicted by price fluctuations.

The rest of the article is presented below. The second section proceeds with a discussion of relevant topics in the related literature. The third section presents the methodological background contains a brief overview of RL methods. The proposed DRL technique is described in the fourth section. The experimental results are illustrated in the fifth section as well as a further discussion about the most interesting outcomes. The paper concludes with conclusions and prospects for future work.

2   Related Works

Since the beginning of trading, traders have sought to maximize the investor's return by using all sorts of methods and techniques including technical analysis [8,11,22] fundamental analysis [28], and algorithmic trading [1,9,27,31], but only some of them have succeeded. Over the past decade, artificial intelligence has attracted increasing interest from the scientific community, changing many areas. The tremendous advantage of deep reinforcement learning methods is a crucial component of this expanding attention. However, little research focuses on the application of artificial intelligence to cryptocurrency trading strategies. Outstanding trading strategies cannot only assist investors to achieve high returns but also help them reduce revenue risk effectively [29]. In this context, some related articles are studies of cryptocurrency trading strategies engaged in developing efficient models for predicting the direction of cryptocurrency prices and then designing specific trading strategies to assist investors in taking advantage of them. Most of them use supervised learning algorithms to predict Bitcoin price movement as [3,10,17,23]. They have applied various Machine Learning (ML) algorithms likewise random forests, Naive Bayes, and logistic regression to obtain a high accuracy of around 98.70 %. With the advent of DL algorithms and their notable success in several fields, notably in image processing [2,4,16] have benefited from this eventful realization to obtain good results based on Deep Neural Networks (DNNs). In the same way, the most recent research [5,36,38] focuses on the application of DL to financial markets to find the best results. Although DL techniques have gained considerable performance in this area, they have many shortcomings, including computational and numerical limitations, sub-optimal optimization of parameters, and poor generalization [37]. More recently, the authors in [15] provided an MDP model adapted to the financial trading problem using the Deep Recurrent Q-Network (DRQN) algorithm that achieves favorable outcomes in many of the simulation parameters [29]. According to these techniques, several studies have already investigated the Reinforcement Learning methods to tackle this algorithmic trading problem with high-quality results. For instance, Azhikodan et al. in [6] aim to validate that RL can successfully learn the tricks of financial trading. Similarly, Hegazy and Mumford [14], in their comparative study showed the best performance of the Recurrent Reinforcement Learning algorithm (RRL). Despite the successes of RL [7], these findings are unable to handle high-dimensional issues due to the lack of scalability [19,26]. This work provides a more rigorous novel framework for evaluating trading effectiveness on DRL, the newest AI technology with great power to tackle the constraints of existing DL approaches. in another effort to accurately evaluate the effectiveness of investment strategies. DRL interacts with the task sequentially, allowing it to progressively understand the task's intrinsic knowledge and then accomplish it efficiently. This family of methods focuses on the learning procedure of an intelligent agent (i) addressing difficult sequential decision-making problems. (ii) interpreting sequentially with an unknown environment (iii) optimizing its cumulative rewards by applying deep learning approaches to generalize the feedback obtained by interacting with the environment [32]. (iv) Processing complex, nonlinear models of economic data effectively surmounts these challenges [13]. DRL brings a novel and powerful AI-based guideline that is getting more popular in autonomous decision-making and operation control by combining the power of DL in solving most of the large and challenging problems with RL, which has a generic, flexible, and exhaustive facility addressing the problems of sequential decision-making. To solve the aforementioned issues, this study proposes a new rule-based policy based on DRL by implementing Dueling Double Deep Q-Networks, Proximal Policy Optimization, and Advantage Actor-Critic to achieve an optimal policy. For more details, a brief overview of RL and DRL is introduced in the following section.

3 Methodological Backgrounds

A brief description of what DRL is along with some of the most popular algorithms used in bitcoin trading will be presented in this section.

3.1 Reinforcement Learning

Reinforcement learning is a special kind of ML that employs a trial-and-error method of learning. This differs from other machines in that RL is a powerful learning strategy that does not require labeled data. The topic has seen a recent renaissance with the successful demonstration of the use of deep learning in RL to solve real-world tasks, where DNNs are employed to make decisions, as mentioned above DRL is a combination of DL and RL and it is the focus of this work. Researchers have recently used DRL and shown success in many studies, especially, in the stock markets to make portfolio bets, they have outperformed traditional cryptocurrencies buying and holding strategies [18]. RL involves formalizing an optimal strategy for ensuring the maximization of the expected cumulative benefit. When interacting in the environment, the RL agent gains rewards and observations. When the trading agent is in state s_t from the set of feasible states S, it must select an action a_t from the set of controllable actions A in at each discrete timestamp t, i.e., sell or buy the considered asset or preserve the current position. The trading system is based upon the decision rule described in Eq. 1, determining which actions to take is based on a policy π which describes the agent's behavior and tells it which actions to select for every available state. As an outcome of every action $a_t \in A$, the agent receives a scalar reward r_t , and the next state $s_{t+1} \in S$ is observed. The probability of transition of each feasible next state s_{t+1} is made to be $Pr(s_{t+1}|s_t, a_t)$, where s_{t+1}. Likewise, each feasible reward r_t'reward probability is given as $Pr(r_t|s_t, a_t)$. Therefore, the expected reward, r_t, received by acting a in current state s is computed by using $\mathbb{E}_{Pr(r_t,s_t,a_t)}(r_t|s_t = s, a_t = a)$.

$$Action \ (A) = \begin{cases} 2, \ if \ \mathbb{E}[pr_{t+n}] > pr_t \\ 1, \ if \ \mathbb{E}[pr_{t+n}] < pr_t \\ 0, \ if \ \mathbb{E}[pr_{t+n}] = pr_t \end{cases} \qquad (1)$$

According to the asset prices, pr_t at time t, the automated trading process chooses an action 1 (i.e. Sell) when the estimated price at time t+n (i.e. $\mathbb{E}[pr_{t+n}]$) is lower than pr_t, and when is greater than pr_t the agent takes an action 2 corresponding to the sell, otherwise it does nothing do (hold, i.e. 0). The number n is referred to as the future time steps.

Two other orders that are widely used in trading techniques are take-profit and stop-loss. When an asset reaches a certain price, both of them are used to buy or sell it. Stop-loss orders are intended to limit the amount of money that can be lost. Take-profit is a strategy for ensuring a profit. The DRL process is illustrated in Fig. 1, and this framework can be modeled as finite MDPs.

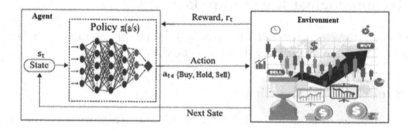

Fig. 1. The DRL framework

The learning agent's goal is to determine an appropriate strategy π^*, which describes the likelihood of acting in state s_t to optimize the total of discounted rewards over time. The formula for calculating the estimated discounted reward G at time t is given as:

$$G_t = \sum_{k=t+1}^{T} \gamma^{k-t-1} r_k \qquad (2)$$

where T is the planning horizon, it can be ∞, and the discounting rate is $\gamma\epsilon$ [0, 1], it identifies the relevance of future returns while emphasizing the importance of recent returns. A greater discount rate value in algorithmic trading indicates that the agent is more long-term investment-oriented. In the case of $\gamma = 0$. Only the immediate reward has an effect. The agent analyzes each reward fairly during the trade in the ultimate scenario of $\gamma = 1$. If we let $\gamma \to 1$, future benefits are taken into account more heavily. This important parameter is modified to attain the desired outcome. Each state that follows the strategy π is assigned a value by the state-value function, $V_\pi : S \to \mathbb{R}$. When the agent starts in state s_t and thereafter following strategy π. The $V\pi$ (s_t) is the expected return value. For MDPs, we can formally define $V_\pi(s_t)$ as.

$$V_\pi(s_t) = \mathbb{E}_\pi [G_t|\ S_t] = \mathbb{E}_\pi \left[\sum_{k=t+1}^{T} \gamma^{k-t-1} r_k \mid S_t \right] \qquad (3)$$

In the same way, when the environment starts at s_t, takes action a_t, and follows policy π, the action-value function, $Q_\pi : S \times A \to \mathbb{R}$ indicated $Q_\pi(st, at)$, formally defined by Eq. (3), offers the expected return.

$$Q_\pi(s_t) = \mathbb{E}_\pi [G_t|\ S_t,\ A_t] = \mathbb{E}_\pi \left[\sum_{k=t+1}^{T} \gamma^{k-t-1} r_k \mid S_t,\ A_t \right] \qquad (4)$$

The RL's goal is to identify a policy that generates a large amount of payoff over time. We can exactly define an optimal strategy for finite MDPs in the following manner. A partial ordering of policies is defined by value functions. If a policy's expected return π is larger than or equal to π_0 for all states, we call it an optimal policy. There can be several of them, however, we consider all of them to be optimal policies. They have the same state-value function, which we refer to as the optimal state-value function and is symbolized by V* as follows:

$$V_\pi{}^*(s_t) = \max_\pi\ V_\pi(s_t),\ \text{for all } s_t \in S.$$

Similarly, the optimal action-value function denoted Q_π^*, is as shown in:

$$Q_\pi{}^*(s_t, a_t) = \max_\pi\ Q_\pi(s_t, a_t)$$

The action-value function can be redefined according to Bellman's optimality equation as follow:

$$Q_\pi{}^* (s_t, a_t) = r + \gamma \max_{a_{t+1}} Q_\pi (s_{t+1}, a_{t+1}) \qquad (5)$$

the Q-learning algorithm is One of the most commonly utilized value-based RL frameworks [34], the agent updates the Q-value of the current state s_t considering the optimal Q-value of the next state $Q_\pi{}^* (s_{t+1}, a_{t+1})$ where action a_{t+1} is a greedy action based on a Q-matrix that is updated at each time step according to Bellman's optimality equation as shown in the following equation:

$$Q_{\pi_{new}}(s_t, a_t) = (1 - \alpha)Q_{\pi_{old}}(s_t, a_t) + \alpha \left[r + \gamma \max_{a_{t+1} \in A} Q_\pi^*(s_{t+1}, a_{t+1}) \right] \quad (6)$$

Under the assumptions that: the state-action pairings are typically described discretely, as well as any actions are repetitively sampled throughout the states to guarantee enough exploration, and thus not required access to the transition model, one generic demonstration of convergence to the optimal value function is known [34]. A $Q_\pi(s, a, \theta)$ function of parametric values is required in this case, with θ denoting the parameters that define the Q-values. In the literature, there are several different Q-networks like DQNs [25] and Double DQNs). The Dueling D-DQNs are used in the proposed trading framework [33], motivated by the successful outcome in [20], the DD-DQNs estimate the value function $V_\pi(s)$ and the related advantage function $A_\pi(s_t, a_t) = Q_\pi(s_t, a_t) - V_\pi(s_t)$ usig a dueling network, thereafter combining them to approximate $Q_\pi(s, a)$. In this model, a two stream of fully connected layers (FC) are succeeded by a convolutional neural network (CNN) layer, that are employed to independently approximate the value and advantage functions before integrating both streams to estimate the action-value function.

In the next section, we explain the RL algorithms applied in this study, DD-DQN, and both advanced policy gradient approaches are employed as A2C and PPO methods.

4   Proposed Deep Reinforcement Learning Module

DD-DQN is the foundation of the planned q-learning trading system. An agent interacts with the financial market in this situation. The agent determines the type of action to take on a bitcoin unit based on the current status of the financial market. After acquiring a bitcoin, it is stored in a portfolio. A take-profit denoted by K_t ($K_t = +15$, and stop-loss denoted by L_s ($L_s = 5\%$)are indeed implemented to the portfolio. For example, if the portfolio loses more than a certain amount (for example, 5%), all open positions are closed. For DRL approaches, the exploration-exploitation issue is critical. Exploitation is concerned with environmental information (i.e. transition and reward functions), whereas exploitation is concerned with maximizing profit. As a result, the agent can perform a random action with probability,, and then follow the optimal policy with probability, $1 - \epsilon$ (ϵ−greedy technique).

4.1   Advantage Actor-Critic

The Asynchronous Actor-Critic (A3C) policy gradient technique has an asynchronous version called the Advantage Actor-Critic policy gradient method [24]. The A2C protocol is used to accelerate policy gradient updates, and minimize the gradient variance of the policy through an advantage function. The model becomes more strong by feeding the quality of action and how it might be improved into the action evaluation phase.

In comparison to the asynchronous design of A3C, and synchronously changes parameters to increase GPU efficiency [24]. It is better able to adapt to risk [35]. A2C is a deterministic, synchronous implementation that allows for each actor to complete their section of experience before adjusting and aggregating across all actors. Because of the bigger batch sizes, GPUs are used more effectively. The A2C update is determined using the following formula:

$$\nabla_\theta J(\theta) = \mathbb{E}[\sum_{t=1}^{T} \nabla_\theta \log \pi_\theta \left(a_t \mid s_t\right) A_\pi \left(s_t, a_t\right)] \tag{7}$$

where the estimate of the advantage function denoted by $A_\pi \left(s_t, a_t\right)$, given by $A_\pi(s_t, a_t) = r(s_t, a_t, s_{t+1}) + \gamma(V_\pi(s_{t+1}) - V_\pi(s_t))$.

4.2 Proximal Policy Gradient

PPO is a new class of policy gradient techniques for RL, that alternates among sampling data via environment interaction and optimizing a clipped "surrogate" objective function L^{Clip} [30], representing the difference of the new policy following the most recent update $\pi_\theta(a \mid s)$ and the previous policy following the latest $\pi_{\theta_k}(a \mid s)$. Traditional policy gradient algorithms only do one gradient update per data sample, whereas PPO has an objective function that allows for several epochs of minibatch updates. PPO outperforms other online policy gradient approaches because it strikes a good balance of sample complexity, simplicity, and wall-time. It has some of the features of trust region policy optimization, but they are far simpler to implement, more generic and have superior sample complexity. The PPO update is determined using the following formula:

$$L\left(s, a, \theta_k, \theta\right) = \min \left(\frac{\pi_\theta \left(a_t, s_t\right)}{\pi_{\theta_k} \left(a_t, s_t\right)} A_{\pi_{\theta_k}} \left(s_t, a_t\right), \text{Clip} \left(\frac{\pi_\theta \left(a_t, s_t\right)}{\pi_{\theta_k} \left(a_t, s_t\right)}, 1 - \epsilon, 1 + \epsilon \right) A_{\pi_{\theta_k}} \left(s_t, a_t\right) \right) \tag{8}$$

where ϵ refers to a constant hyperparameter.

5 Experimental Results

The proposed trading systems are tested on bitcoin historical data on the platform www.bitcoincharts.com and www.kaggle.com. Specifically, we consider the price of bitcoin in USD from December 1, 2016, to December 19, 2020. The observation space is collected using open, high, low, and close prices. The in-sample period and the out-of-sample period are both taken into account. The first section offers information on the training and validation phases. The second section offers information on the trade stage. DD- DQN, PPO, and A2C are used to learn three agents in the training stage. Then comes the validation stage, which involves validating the three agents using the Sharpe ratio (Sh) presented in Eq. (9) and changing critical parameters like learning rate, number of episodes, and so on. Finally, we analyze the profitability of each algorithm throughout the trading stage.

$$Sh = \frac{\bar{r}_r - r_f}{\sigma_s} \tag{9}$$

where $\bar{r}_r$ indicates the expected asset return, the r_f is the risk-free factor, and σ_s stands for the asset's standard deviation. In our validation stage, we also use the volatility metric to adjust risk aversion. A simple profit function pr_{profit} is also used to improve the performance of our outcomes is determined by subtracting the asset price value at time t (pr_t) from its at time t−1 value (pr_{t-1}).

The proposed algorithms are tested using the Colab GPU and Python implementation.

Table 1, illustrated the proposed results during 12 quarters (Quart), since janvier, 2018 to december, 2020. As shown in Table 1, DD-DQNs have the greatest validation Sharpe ratio Sh = 0.05 from 2017/10 to 2017/12, and we choose it to trade for the next period between 2018/01 to 2018/03 (Quart1). And we selected it because it has the best value of Sh = 0.63 from the first quarter. We also use the second quarter trade(i.e. 2018/04–2018/06). We can see that the A2C agent has the best validation Sh = 0.2 from 2018/04 to 2016/06 (Quart2), then for the following quarter through 2018/07 to 2018/09, we can choose A2C for trading. Trading for the following period (Quart4) from 2018/10 to 2018/12, we select PPO which has the good validation Sh = −0.15 from 2018/07 to 2018/09. And so on.

Table 1. The value of Sharpe ratios under time

Quarterly trading	DD-DQNs	PPO	A2C	Chosen model
Quart 1	**0.05**	0.04	0.02	DD-DQNs
Quart 2	**0.63**	0.46	0.50	DD-DQNs
Quart 3	0.09	0.02	**0.2**	A2C
Quart 4	−0.11	**0.22**	−0.59	PPO
Quart 5	0.38	**0.49**	0.28	PPO
Quart 6	**0.36**	0.25	0.36	DD-DQNs
Quart 7	**0.09**	0.01	0.04	DD-DQNs
Quart 8	0.09	0.02	**0.2**	PPO
Quart 9	0.36	0.46	**0.51**	A2C
Quart 10	0.16	−0.28	**0.36**	A2C
Quart 11	0.10	0.09	**0.13**	A2C
Quart 12	0.08	0.08	**0.16**	A2C

The average return and Sharpe Ratio for different agent RL used in this work are calculated to compare the agent Performance. The result is illustrated in the following table.

Table 2. Performance evaluation comparison

(2018/01–2020/12)	Proposed method	DD-DQNs	PPO	A2C
Profit reward function (average return (%))	12.84	12.6	11.75	09.84
Sharpe ratio	1.41	1.39	1.25	1.36

We can observe from the results presented in Table 2 that the DD-DQNs agent is the best at tracking trends and generates better yields, among the PPO and A2C agents, it has the greatest average yields, at 12.6%. When faced with a rising market, the DD-DQNs are favored. In a bullish market, the PPO method performs similarly to DD-DQNs, but not as well. It can be used as a supplementary to DD-DQNs. The proposed method is based on a combination of the three agents, it is clear that the average return of the proposed approach reaches a higher value than other agents.

As shown in Table 2, the aggregate strategy has a Sharpe ratio of 1.41, outperforming PPO, A2C, and DD-DQN, which have Sharpe ratios of 1.25, 1.36, and 1.39, respectively. As a result, the proposed aggregate policy can be used to establish a trading policy that surpasses the three individual agents.

6   Conclusion

In this study, we investigated the power of applying the DD-DQNs, A2C, and PPO algorithms to learn a cryptocurrency trading strategy. Over roughly two years, these strategies are compared. Based on profit reward functions, and the Sharpe ratio to evaluate the performance of different methods. The proposed approach is made by combining the previous approaches. The evaluation framework showed that our technique is the most profitable approach for cryptocurrency trading.

Future work is attractive for investigating more advanced models, dealing with larger-scale data, we can introduce other features (adding advanced transaction costs, liquidity mode, ...) for the state space.

References

1. Ahmad, I., Ahmad, M.O., Alqarni, M.A., Almazroi, A.A., Khalil, M.I.: Using algorithmic trading to analyze short term profitability of bitcoin. Peer J. Comput. Sci. **7**, e337 (2021)
2. El Akraoui, B., Daoui, C.: Deep learning for medical image segmentation. In: Fakir, M., Baslam, M., El Ayachi, R. (eds.) CBI 2021. LNBIP, vol. 416, pp. 294–303. Springer, Cham (2021). https://doi.org/10.1007/978-3-030-76508-8_21
3. Alessandretti, L., ElBahrawy, A., Aiello, L.M., Baronchelli, A.: Anticipating cryptocurrency prices using machine learning. Complexity **2018**, 1 (2018)

4. Arévalo, A., Niño, J., Hernández, G., Sandoval, J.: High-frequency trading strategy based on deep neural networks. In: Huang, D.-S., Han, K., Hussain, A. (eds.) ICIC 2016. LNCS (LNAI), vol. 9773, pp. 424–436. Springer, Cham (2016). https://doi.org/10.1007/978-3-319-42297-8_40
5. Murillo, A.R.A.: High-frequency trading strategy based on deep neural networks. In: Ingeniería de Sistemas (2019)
6. Azhikodan, A.R., Bhat, A.G.K., Jadhav, M.V.: Stock trading bot using deep reinforcement learning. In: Saini, H.S., Sayal, R., Govardhan, A., Buyya, R. (eds.) Innovations in Computer Science and Engineering. LNNS, vol. 32, pp. 41–49. Springer, Singapore (2019). https://doi.org/10.1007/978-981-10-8201-6_5
7. Britz, D.: Introduction to learning to trade with reinforcement learning (2018). https://www.wildml.com/2018/02/introduction-to-learning-to-tradewith-reinforcement-learning
8. Chaboud, A.P., Chiquoine, B., Hjalmarsson, E., Vega, C.: Rise of the machines: algorithmic trading in the foreign exchange market. J. Finan. **69**(5), 2045–2084 (2014)
9. Chan, E.P.: Quantitative Trading: How To Build Your Own Algorithmic Trading Business. John Wiley and Sons, New York (2021)
10. Colianni, S., Rosales, S., Signorotti, M.: Algorithmic trading of cryptocurrency based on twitter sentiment analysis. CS229 Project **1**(5), 1–4 (2015)
11. Dash, R., Dash, P.K.: A hybrid stock trading framework integrating technical analysis with machine learning techniques. J. Finan. Data Sci. **2**(1), 42–57 (2016)
12. Goldkamp, J., Dehghanimohammadabadi, M.: Evolutionary multi-objective optimization for multivariate pairs trading. Expert Syst. Appl. **135**, 113–128 (2019)
13. Guo, Y., Fu, X., Shi, Y., Liu, M.: Robust log-optimal strategy with reinforcement learning. arXiv preprint arXiv:1805.00205 (2018)
14. Hegazy, K., Mumford, S.: Comparitive automated bitcoin trading strategies. CS229 Project **27**, 1–6 (2016)
15. Huang, C.Y.: Financial trading as a game: a deep reinforcement learning approach. arXiv preprint arXiv:1807.02787 (2018)
16. Isensee, F., Jaeger, P.F., Simon, A.A., Kohl, J.P., Maier-Hein, K.H.: NNU-net: a self-configuring method for deep learning-based biomedical image segmentation. Nat. Methods **18**(2), 203–211 (2021)
17. Jang, H., Lee, J.: An empirical study on modeling and prediction of bitcoin prices with Bayesian neural networks based on blockchain information. IEEE Access **6**, 5427–5437 (2017)
18. Jiang, Z., Liang, J.: Cryptocurrency portfolio management with deep reinforcement learning. In: 2017 Intelligent Systems Conference (IntelliSys), pp. 905–913. IEEE (2017)
19. Kim, Y., Ahn, W., Oh, K.J., Enke, D.: An intelligent hybrid trading system for discovering trading rules for the futures market using rough sets and genetic algorithms. Appl. Soft Comput. **55**, 127–140 (2017)
20. Lucarelli, G., Borrotti, M.: A deep reinforcement learning approach for automated cryptocurrency trading. In: MacIntyre, J., Maglogiannis, I., Iliadis, L., Pimenidis, E. (eds.) AIAI 2019. IAICT, vol. 559, pp. 247–258. Springer, Cham (2019). https://doi.org/10.1007/978-3-030-19823-7_20
21. Ma, W., Xuemin, T., Luo, B., Wang, G.: Semantic clustering based deduction learning for image recognition and classification. Pattern Recogn. **124**, 108440 (2022)
22. Markowitz, H.: Portfolio selection in the journal of finance, vol. 7 (1952)

23. McNally, S., Roche, J., Caton, S.: Predicting the price of bitcoin using machine learning. In: 2018 26th EuroMicro International Conference On Parallel, Distributed and Network-based Processing (PDP), pp. 339–343. IEEE (2018)
24. Mnih, V., et al.: Asynchronous methods for deep reinforcement learning. In: International Conference on Machine Learning, pp. 1928–1937. PMLR (2016)
25. Mnih, V., et al.: Playing Atari with deep reinforcement learning. arXiv preprint arXiv:1312.5602 (2013)
26. Mosavi, A., et al.: Comprehensive review of deep reinforcement learning methods and applications in economics. Mathematics 8(10), 1640 ((2020))
27. Murphy, J.J.: Technical Analysis of The Financial Markets: A Comprehensive Guide To Trading Methods and Applications. Penguin (1999)
28. ÖZYEŞİL, M.: Comparison of technical and fundamental analysis trading disciplines on portfolio performance: short and long term backtest analysis on Borsa Istanbul national stock indices. J. Contempor. Res. Bus. Econ. Finan. 3(3), 128–143 (2021)
29. Sattarov, O., et al.: Recommending cryptocurrency trading points with deep reinforcement learning approach. Appl. Sci. 10(4), 1506 (2020)
30. Schulman, J., Wolski, F., Dhariwal, P., Radford, A., Klimov, O.: Proximal policy optimization algorithms. arXiv preprint arXiv:1707.06347 (2017)
31. Schwager, J.D.: A Complete Guide To The Futures Market: Technical Analysis, Trading Systems, Fundamental Analysis, Options, Spreads, and Trading Principles. John Wiley & Sons, New York (2017)
32. Théate, T., Ernst, D.: An application of deep reinforcement learning to algorithmic trading. Expert Syst. App. 173, 114632 (2021)
33. Wang, Z., Schaul, T., Hessel, M., Hasselt, H., Lanctot, M., Freitas, N.: Dueling network architectures for deep reinforcement learning. In: International Conference on Machine Learning, pp. 1995–2003. PMLR (2016)
34. Watkins, C.J., Dayan, P.: Q-learning. Mach. Learn. 8(3), 279–292 (1992)
35. Yang, H., Liu, X.-Y., Zhong, S., Walid, A.: Deep reinforcement learning for automated stock trading: an ensemble strategy. In: Proceedings of the First ACM International Conference on AI in Finance, pp. 1–8 (2020)
36. Yu, P., Yan, X.: Stock price prediction based on deep neural networks. Neural Comput. App. 32(6), 1609–1628 (2019). https://doi.org/10.1007/s00521-019-04212-x
37. Zhang, D.: Deep reinforcement learning in medical object detection and segmentation (2020)
38. Zhang, Z., Zohren, S., Roberts, S.: Deep learning for portfolio optimization. J. Finan. Data Sci. 2(4), 8–20 (2020)

An Exploration of Student Grade Retention Prediction Using Machine Learning Algorithms

Aomar Ibourk[1] and Ismail Ouaadi[2(✉)]

[1] Microeconometrics, LARESSGD, University of Marrakech, Marrakech, Morocco
[2] Management Sciences, Faculty of Legal, Economic and Social Sciences Agdal,
Mohammed V University of Rabat, Rabat, Morocco
ismail.ouaadi@gmail.com

Abstract. Education is an important determinant of nation that succeed, nowadays, artificial intelligence algorithms have been applying practically in all field of science. As a result, learning analytics was born, referring to artificial intelligence techniques used in the field of education. The aim of this work is to build a machine learning models that can predict the student grade retention in K-9 grade. This work, first, applies six supervised artificial intelligence techniques, and validate them within four scenarios according the normalization and balancing of data in a second step, finally, these models was validated according to four recognized measures in order to choose the one that fit very well. The final purpose of our work is to contribute to the education field, we achieve this by providing some new idea where we have applied machine learning algorithms in grade retention research issue.

Keywords: Grade retention · Prediction · Machine learning algorithms

1 Introduction

Grade retention have got more intention in academic research regarding its importance for the community and considering the biggest governmental investments in the education domain. In fact, the Moroccan government does not escape this observation, and the education investment constitutes the first budget consumption with up to the 17% of national budget regard to payment appropriations (variation from 2018 to 2019) [1].

Moreover, the grade retention has a negative impact on the students themselves and either their parents on life aspects like financial and social ones. Furthermore, the grade retention has a bad effect on the most players in education from teacher to school staff and in at all education level from early school to the college.

This issue is omnipresent in all countries, developed or not, which spend more resource and time in attention to succeed the related policies. Moreover, class retention has spilled a lot of ink given the huge number of scientific studies, that attempt to explain this topic and try to find some solutions to mitigate it.

In this perspective our paper was conducted, by developing models which are based on some determinants of grade retention that we found significant in a first step, and

M. Fakir et al. (Eds.): CBI 2022, LNBIP 449, pp. 94–106, 2022.
https://doi.org/10.1007/978-3-031-06458-6_8

by introducing machine learning algorithms to train these models to figure out the best ones that allow predicting student grade retention. To do so our work is organized as follow. The Sect. 2, namely, literature review gives some works in this field, while the Sect. 3 describes the methodology followed in our contribution, Sect. 4 presents our experiment and discuss our results and the final section provides some conclusion and future research path.

2 Literature Review

For the above reasons, many works have dealt with this problem to make clear explanations and to find some solutions. In [2] the authors analyze the factors that are associated with the probability for a student to be retained in primary school, they targeted the European countries and the data was gathered from the PIRLS 2011 results. They used the multiple regression method and two categories of features, features related to the student and others related to the school. They found that there are common factors that make a student more/less at-risk to be retained. These factors, are classified in four groups: (i) family's socioeconomic background, (ii) attending pre-school, (iii) available (economic and cultural) resources, and (iv) student composition of the school attended.

The second work is that of [3] it analyses the school enrollment and graduation rates in China using the 1990 Chinese Census, the authors have targeted the youth aged student from 10–18. They used the regression analysis to find determinants of educational in rural and urban context.

Other studies have focused on specific topics in education like self-efficacy [4], where the author tries to explore the profiles of self-efficacy and its associations with gender, socioeconomic status, fear of failure and academic achievements among students from different countries.

Many other works have focused on different determinants and factors that have an impact on grade retention and they used different machine learning technics such as Logistic regression [5], K-Nearest Neighbors, Decision Tree, Random Forest and Support Vector Machines and Neural Networks [6–10] which is our aim in this work.

3 Methodology

The development of this section attempts to discuss the methodology followed in this work. Our methodology takes into account the highlights provided in [6], which describe the development of predictive learning analytic models that is composed of four components to be considered in studies:

– Data sources and student variables;
– Data preprocessing and handling;
– Machine learning techniques; and
– Evaluation of accuracy and generalizability.

Given these components this section gives, in first, a brief description of dataset and used features, and secondly an overview of data preprocessing, where we present how we have dealing with aspects related to data like missing value, features selection, balancing datasets, discretization of variables, variable normalization and data splitting to train and test sets. The end of this section explains the predictive models chosen to be performed. Moreover, the fourth point was cover in the next section.

3.1 Data Description

The dataset considered in this work consists of the student K-9 records for the Moroccan intermediate school. The data is collected from different regions of Morocco, in an intention of the diversification of data, and with fully respect of the ethical guidelines and laws. The dataset is composed of 6814 records and 40 features at all, given the subject of this research, which is about student class retention prediction, we have selected seven following features which are related to the students and their environment, READ_score, MATH_score, SCIE_score, Gender, Mother_edu, School_type and Repeat as described in the following Table 1.

The Repeat variable is considered as the target, which its values are Boolean, True means that the student has repeated in a class and False otherwise. According to the target population, student of the final grade of the middle school (K-9), the dataset has reduced to 2221 records.

Table 1. The list of student features with description.

Feature	Description	Values
Gender	Gender of the student	Male, Female
READ_score	Score of reading	Real
MATH_score	Score of mathematics	Real
SCIE_score	Score of science	Real
Mother_edu	Mother's level of education	From 1 to 4
School_type	School type derived from sampling information	Public, private
Repeat	Grade Repetition	True, False

These dataset records are recovered by a questionnaire and the final test score of each student of three test in courses related to the middle school in Morocco, these scores are determined based on the international test PISA. The MATH_score corresponds to PISA assessment ranking for each student, and the same for the other features related to the science and reading. Where the mother's level of education is filled according to the International Standard Classification of Education (ISCED), and school type public signifies the governmental school, and private school means proprietary school. The descriptive statistics of our dataset, is given in the Table 2 below.

Table 2. Descriptive statistics (a) continuous variables

	READ_score	MATH_score	SCIE_score
frequency	2221	2221	2221
mean	333.316	342.546	352.065
std	47.900	46.162	41.766
min	184.101	183.583	200.072
25%	298.817	311.260	324.105
50%	330.018	342.182	349.507
75%	363.852	372.141	378.208
max	506.499	520.142	605.104

(b) ordinal variables

	Gender	Repeat	Mother_edu	School_type
frequency	2221	2221	2221	2221
std	0.498	0.365	1.289	0.186
min	1	0	1	1
25%	1	1	1	2
Median (50%)	2	1	2	2
75%	2	1	4	2
max	2	1	4	2

The features selection and their importance are based on the literature review [2, 3, 10–15] and rather they are chosen according to the student's privacy [9] requirements. The Moroccan laws imposes ethical and privacy restrictions that we should respect, for these raisons we have removed irrelevant features, like student name, student's ID. Other features are ignored because they are considered endogenous ones, for example the age and other courses criteria. Furthermore, the correlation coefficients (KANDAL, SPEARMAN and PEARSON) are calculated and taken into consideration to select the relevant feature whatever their correlation sense, these correlation coefficients, with the retention in grade feature, are provided in the next Table 3.

Table 3. Correlation coefficient of the repeat variable

	KANDALL	SPEARMAN	PEARSON
Gender	0.076598	0.076598	0.076598
School_type	0.068605	0.068605	0.068605
Mother_edu	0.016223	0.017192	0.006311
READ_score	−0.038804	−0.047514	−0.07281
MATH_score	−0.05715	−0.069978	−0.086276

3.2 Data Preprocessing

The preprocessing dataset is a crucial step, it determines the performance of the data analytics, and should be taken carefully as mentioned in [6, 11]. The training step of machine learning algorithm is the most sensitive one for data cleaning, for this reason all criteria related to this purpose are answered as in the following subsections. While the dataset extraction and preparation tasks are given in Fig. 1, the pre-processing tasks are schematized in the following Fig. 1.

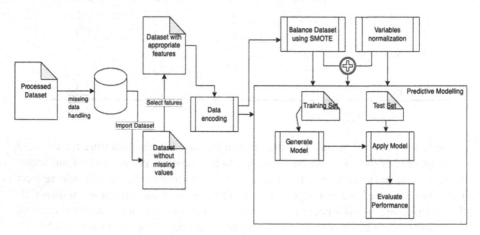

Fig. 1. Pre-processing tasks performed in Pandas

Missing Value: the missing data represent around 4% localized in two features (Mother's level of education and type of school) which represent 110 missed values for each one in our Database. Many technics are used to deal with missing data [2], like deleting corresponding records or replacing the missing values, the latest choice was chosen in our case due to the irrelevant rate of missing data. The missed values are filled with the mode of specified features, which consists of the value that mostly appears in categorical data, or their means as described in [16].

Feature Selection: as we mentioned above, the feature selection step was conducted according to the student's privacy and Moroccan laws about the access to the private information. Indeed, we have excluded features raising these issues like age and other ones related to the student like IDs, names, while the score of other courses is considered as endogen features and other related to his environment like school name. It has many technics that can be used for feature selection like gain ration (GR) [9], but we have opted for the correlation method, which involves looking for relationships between variables, to determine the importance of our feature as described in Table 3, also used by [17]. Finally, our dataset contains seven features, one target and six predictors, and 2221 student records.

Balancing Datasets: many works have highlighted the pertinence of the balancing datasets and show the raised performance of machine learning technics after introducing the SMOTE method [10, 11] for balancing datasets. In our work we have checked the relevance of the SMOTE method applied to our data by comparing the final evaluation of machine learning algorithms that we have performed before and after balancing.

Variable Normalization: in the same way of balancing datasets, we analyze the effect of the data normalization as recommended by [6]. In fact, the effect analyzing is performed by applying variable normalization before and after SMOTE. The MINMAX method was chosen for this purpose as processed in [18]. The importance of data normalization in machine learning performance has shown in the [12].

Training and Validation: to make it, we proceed to split our dataset into training data to train models and testing data for validating the models. To do so, we shuffle data firstly and sliced them in a second time within the train_test_split function of scikit-learn python framework. The result is 80% for data training and the 20% remained for data testing as recommended in [18].

3.3   Predictive Modelling

In this work we have chosen six algorithms of machine learning for classification purpose, namely, Logistic Regression, k-nearest neighbors, Decision Tree, Random Forest, Support vector machines and Artificial Neural Network. These algorithms are performed for a classification into two classes of the target variable (Repeat) where the student has a retention previously (YES = 1) or not (NO = 0) repeated a grade during their school career.

Logistic Regression (LR): Is one of the well-known techniques in both machine learning and statistics [8], it can be in the form of Eq. (1). LR is deployed in many binary classification cases, and it has used in the works [3, 8]. In our work we utilize the LR to predict student retention in forthcoming years of school using the predictors discussed above. The test score and validation measures are given in the next section.

$$L_{(f(x))} = \frac{1}{1 + e^{-f(x)}}$$ (1)

K-Nearest Neighbors (KNN): is the simplest machine learning technique that can be implemented. The KNN is implemented in many works [7, 8, 10], it allows the classification of a given data point based on its k-nearest neighborhoods. The distance between the two data points is measured using the Manhattan distance (Eq. 2) or Euclidian distance (Eq. 3) [10].

$$d(x, y) = \|x - y\| = \sum_{i=1}^{n} |x - y|$$ (2)

$$d(x, y) = \|x - y\| = \sqrt{\sum_{i=0}^{n} (x_0 - y_0)^2}$$ (3)

where (x, y) are vectors.

Decision Tree (DT): is another popular technique of machine learning that consists of a classification model that has the form of a tree structure [8], where the internal nodes contain the independent features and the leaves are destined for the possible targets. The DT technique can be used for either classification or regression problems. In our case it used for the classification issue, and it allows to understand the importance of each predictor unlike artificial neural networks.

Random Forest (RF): the main idea of this method is to create different decision trees, where each DT contains a randomly chosen subset of the independent variable [8]. Thus, it is a technique based on decision tree method, that gives us a variety of DTs with different structures for solving the same problem. It has many other advantages besides those related to the Decision Tree nearly discussed.

Support Vector Machines (SVM): also called kernelized support vector machines, is a technique that allows to deal with complex problems that are not defined by hyper-planes in the input space [18]. They are two kind of SVM algorithms, the classification one and the regression algorithm.

Artificial Neural Network (ANN): is the famous technique of machine learning, and it's the one that put artificial intelligence back in the spotlight after a long period of oblivion [19]. Nowadays, it constitutes the revolutionary technique of machine learning, namely, Deep Learning which is deploys a complex model of neural networks. In this work, we chose a relatively simple methods of ANN, called multilayer perceptron (MLP), it viewed as a generalization of linear models that perform multiple stages of processing to come to a decision [18]. Despite, it is considered as a Blackbox technique, it shows good performances in many works.

4 Experiment and Results

In this section, we present our experiments, report the findings of the training and testing the algorithms, the results are discussed in the final subsection.

4.1 Experiment Settings and Evaluation Measures

As we can see her, all these techniques of machine learning are qualified as supervised techniques. As described in Fig. 1, we have performed these six techniques in three times. The first round, we train and test our models without balancing and normalization of data, secondly, we doing so with the SMOTE technique that allows us balancing data, in the third round we have normalized our dataset with the MINMAX method and without SMOTE. Finally, we have processed our models with normalized data and SMOTE. Results are given in the next subsection.

In fact, the relevance of the developed models should be measured, for this purpose we have use a set of evaluation measures, given as follow.

- Classification accuracy: which evaluates the performance of the prediction models on the whole. It describes how the models correctly identified the True Positive (TP) and True Negative (TN) instances of the confusion Matrix. Its formula (4):

$$accuracy = \frac{(TP + TN)}{(TP + TN + FP + FN)} \tag{4}$$

- Recall and Precision: while the recall refers to completeness of the model the precision refers to the exactness of the model [9]. Their formulas (5) and (6) are the following:

$$Recall = \frac{TP}{(TP + FN)} \tag{5}$$

$$Precision = \frac{TP}{(TP + FP)} \tag{6}$$

- F-Measure: which is the harmonic average of the pcision and recall values, it measures the performance of the model in a single value, and it is calculated as the followed (formula (7)):

$$F - Measure = 2 * \frac{(Precision * Recall)}{(Precision + Recall)} \tag{7}$$

4.2   Results and Discussion

In this subsection, we give the findings of our research and discuss them briefly.

Based on results given in the Tables 4, 5, 6, and 7 we can summarize the performance of the models as generally good in the first round, which consists of processing models before balancing and normalization of data. This round has the highest values in three validation measures (accuracy, recall and f-measure). Where the Decision Tree Classifier model has the greatest values with respectively 0.849, 1 and 0.918 in the all measures. In contrast the SVM Classifier model has the greatest precision score in the second round that consists of balancing data.

The graphical representation of validation measures (Figs. 2, 3, 4 and 5) of the LR model has the best performance in the first round (A) and the second highest score of performance in the third round (C) after the normalization and before SMOTE, and the lowest one in the second round (B) according to the accuracy recall and f-measure measures.

The KNN algorithm has performed very well in the third round (C) with 0.813, 0.854, 0.94 and 0.895 respectively in the accuracy, precision, recall and f-measure. And the lowest scores in the second round. In the same rounds, the RF classifier has got 0.827, 0.863, 0.94 and 0.902 respectively in the accuracy, precision, recall and f-measure in the third round, and 0.778, 0.879, 0.85 and 0.866 respectively in the accuracy, precision, recall and f-measure in the second round. The SVM model has the best performance in the first and third round with the scores of validation measures 0.843, 0.843, 1, 0.915 respectively in the accuracy, precision, recall and f-measure. Finally, MLP has the highest scores measures in both round (second and third) with 0.843, 0.843, 1 and 0.915 respectively in the accuracy, precision, recall and f-measure, and the lowest performance in the second round with 0.728, 0.874, 0.79 and 0.831 respectively in the accuracy, precision, recall and f-measure.

According to the f-measure the performance of the models, in all rounds, is ranked as follow. In the first rank we found LR model followed by MLP in the second rank, the RF classifier third, DT classifier fourth, KNN model in the fifth rank and in the last rank the SVM classifier. Meanwhile, the SVM has the last f-measure score, it has the first rank in the first and third round according to the recall measure.

Table 4. Before SMOTE and normalization

Model	Tp	Tn	Fp	Fn	Accuracy	Precision	Recall	F
LogisticRegression	375	1	69	0	0.845	0.845	1	0.916
KNeighborsClassifier	351	9	61	24	0.809	0.852	0.936	0.892
DecisionTreeClassifier	373	5	65	2	0.849	0.852	0.995	0.918
RandomForestClassifier	351	14	56	24	0.82	0.862	0.936	0.898
SVMClassifier	375	0	70	0	0.843	0.843	1	0.915
RNNClassifier	375	0	70	0	0.843	0.843	1	0.915

Table 5. After SMOTE and before normalization

Model	Tp	Tn	Fp	Fn	Accuracy	Precision	Recall	F
LogisticRegression	340	13	57	35	0.793	0.856	0.907	0.881
KNeighborsClassifier	297	26	44	78	0.726	0.871	0.792	0.83
DecisionTreeClassifier	306	21	49	69	0.735	0.862	0.816	0.838
RandomForestClassifier	320	26	44	55	0.778	0.879	0.853	0.866
SVMClassifier	2	70	0	373	0.162	1	0.005	0.011
RNNClassifier	297	27	43	78	0.728	0.874	0.792	0.831

Table 6. Before SMOTE and after normalization

Model	Tp	Tn	Fp	Fn	Accuracy	Precision	Recall	F
LogisticRegression	375	0	70	0	0.843	0.843	1	0.915
KNeighborsClassifier	352	10	60	23	0.813	0.854	0.939	0.895
DecisionTreeClassifier	353	5	65	22	0.804	0.844	0.941	0.89
RandomForestClassifier	354	14	56	21	0.827	0.863	0.944	0.902
SVMClassifier	375	0	70	0	0.843	0.843	1	0.915
RNNClassifier	375	0	70	0	0.843	0.843	1	0.915

Table 7. After SMOTE and normalization

Model	Tp	Tn	Fp	Fn	Accuracy	Precision	Recall	F
LogisticRegression	349	11	59	26	0.809	0.855	0.931	0.891
KNeighborsClassifier	301	27	43	74	0.737	0.875	0.803	0.837
DecisionTreeClassifier	306	21	49	69	0.735	0.862	0.816	0.838
RandomForestClassifier	317	26	44	58	0.771	0.878	0.845	0.861
SVMClassifier	349	10	60	26	0.807	0.853	0.931	0.89
RNNClassifier	345	9	61	30	0.796	0.85	0.92	0.883

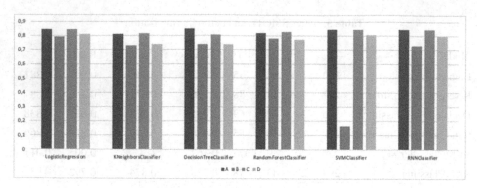

Fig. 2. Accuracy scores of 6 models

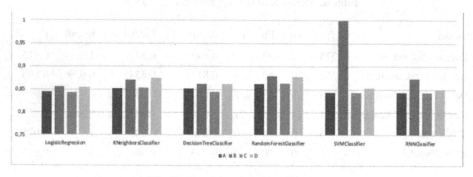

Fig. 3. Recall scores of the 6 models

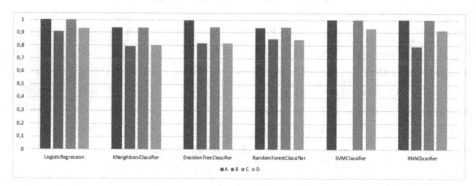

Fig. 4. Precision scores of the 6 models

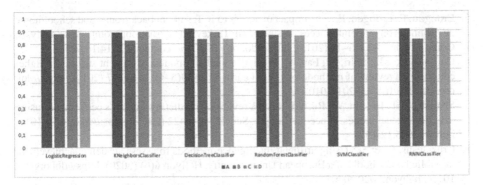

Fig. 5. The F-measure scores of the 6 models

5 Conclusion and Future Work

In this research, we used six models of supervised machine learning to predict the student grade retention for the K-9 grade. The dataset was gathered from the Programme for International Student Assessment (PISA) 2018 Database. After the preprocessing database we get a dataset that contains 2221 student records and seven features.

After analyzing the results related to our model experimentation, we found that Decision tree classifier has the best performance according to the mean of all validation measures and in all rounds. These results allow us to have a clear determination of the importance of the features taken in this research to predict the grade retention of the Moroccan students.

Many works attempt to predict the grade retention in many levels of the educational pathways, this work use different technique to model this subject from those that applied linear regression to the other that applied machine learning algorithm as indicated in the literature review section.

In the following work, we try to model different aspects of retention by introducing new features and new techniques of machine learning, in intention the draw a best comprehension of this subject that consists of a traditional field of education research.

References

1. Royaume Du Maroc Cour Des Comptes: L'exécution du Budget de l'Etat de l'année 2019, December 2020
2. Agasisti, T., Cordero, J.M.: The determinants of repetition rates in Europe: early skills or subsequent parents' help? J. Policy Model. **39**(1), 129–146 (2017). https://doi.org/10.1016/j.jpolmod.2016.07.002
3. Connelly, R., Zheng, Z.: Determinants of school enrollment and completion of 10 to 18 year olds in China. Econ. Educ. Rev. **22**(4), 379–388 (2003). https://doi.org/10.1016/S0272-775 7(02)00058-4
4. Ma, Y.: A cross-cultural study of student self-efficacy profiles and the associated predictors and outcomes using a multigroup latent profile analysis. Stud. Educ. Eval. **71** (2021). https://doi.org/10.1016/j.stueduc.2021.101071

5. Erdogdu, F., Erdogdu, E.: The impact of access to ICT, student background and school/home environment on academic success of students in Turkey: an international comparative analysis. Comput. Educ. **82**, 26–49 (2015). https://doi.org/10.1016/j.compedu.2014.10.023

6. Cui, Y., Chen, F., Shiri, A., Fan, Y.: Predictive analytic models of student success in higher education: a review of methodology. Inf. Learn. Sci. **120**(3–4). 208–227 (2019). https://doi.org/10.1108/ILS-10-2018-0104

7. Shahiri, A.M., Husain, W., Rashid, N.A.: A review on predicting student's performance using data mining techniques. Procedia Comput. Sci. **72**, 414–422 (2015). https://doi.org/10.1016/j.procs.2015.12.157

8. Cruz-Jesus, F., et al.: Using artificial intelligence methods to assess academic achievement in public high schools of a European Union country. Heliyon **6**(6) (2020). https://doi.org/10.1016/j.heliyon.2020.e04081

9. Khan, I., Ahmad, A.R., Jabeur, N., Mahdi, M.N.: An artificial intelligence approach to monitor student performance and devise preventive measures. Smart Learn. Environ. **8**(1), 1–18 (2021). https://doi.org/10.1186/s40561-021-00161-y

10. Zeineddine, H., Braendle, U., Farah, A.: Enhancing prediction of student success: automated machine learning approach. Comput. Electr. Eng. **89** (2021). https://doi.org/10.1016/j.compeleceng.2020.106903

11. Bilquise, G., Abdallah, S., Kobbaey, T.: Predicting student retention among a homogeneous population using data mining. In: Hassanien, A.E., Shaalan, K., Tolba, M.F. (eds.) AISI 2019. AISC, vol. 1058, pp. 35–46. Springer, Cham (2020). https://doi.org/10.1007/978-3-030-31129-2_4

12. Huang, S., Fang, N.: Predicting student academic performance in an engineering dynamics course: a comparison of four types of predictive mathematical models. Comput. Educ. **61**(1), 133–145 (2013). https://doi.org/10.1016/j.compedu.2012.08.015

13. Alodat, M.: Predicting student final score using deep learning. In: Bhatia, S.K., Tiwari, S., Ruidan, S., Trivedi, M.C., Mishra, K.K. (eds.) Advances in Computer, Communication and Computational Sciences. AISC, vol. 1158, pp. 429–436. Springer, Singapore (2021). https://doi.org/10.1007/978-981-15-4409-5_39

14. Cockx, B., Picchio, M., Baert, S.: Modeling the effects of grade retention in high school. J. Appl. Economet. **34**(3), 403–424 (2019). https://doi.org/10.1002/jae.2670

15. Belot, M., Vandenberghe, V.: Evaluating the 'threat' effects of grade repetition: exploiting the 2001 reform by the French-Speaking Community of Belgium. Educ. Econ. **22**(1), 73–89 (2014). https://doi.org/10.1080/09645292.2011.607266

16. Badr, G., Algobail, A., Almutairi, H., Almutery, M.: Predicting students' performance in university courses: a case study and tool in KSU Mathematics Department. Procedia Comput. Sci. **82**, 80–89 (2016). https://doi.org/10.1016/j.procs.2016.04.012

17. Sathe, M.T., Adamuthe, A.C.: Comparative study of supervised algorithms for prediction of students' performance. Int. J. Modern Educ. Comput. Sci. **13**(1), 1–21 (2021). https://doi.org/10.5815/ijmecs.2021.01.01

18. Müller, A.C., Guido, S.: Introduction to Machine Learning with Python: A Guide for Data Scientists, 1st edn. O'Reilly Media Inc., Sebastopol (2016)

19. Russell, S.J., Norvig, P.: Artificial Intelligence: A Modern Approach, 4th edn. Pearson, Hoboken (2021)

Deep Learning Model for Educational Recommender Systems

Abdelkader Grota(✉), Mohammed Erritali, and Elmoufidi Abdelali

Data4Earth Laboratory, Faculty of Sciences and Techniques,
Sultan Moulay Slimane University, Beni Mellal, Morocco
abdelkader.grota@usms.ac.ma

Abstract. Research in the field of recommender systems is evolving rapidly and these systems are increasingly being applied to specific domains, including educational technologies. In the field of education, in general, recommender systems are used to improve the processes of online learning and teaching. However, with the growth in the number of educational resources and their diversities, the problem of information overload is becoming increasingly critical. Therefore, providing learners with personalized educational recommendation tools is a necessity. The objective of this work is to propose a new approach to recommend resources to students according to their preferences. This recommendation approach introduces a general framework called collaborative filtering (CF) based on neural networks, which complements classical models and machine learning algorithms such as KNN, SVD of collaborative filtering. The experiments we have performed prove the performance of the proposed approach.

Keywords: Collaborative filtering · Matrix factorization · Deep learning · Neural networks

1 Introduction

Students are not immune to the problem of infobesity or information overload. A master's or doctoral student, for example, is faced with a huge stream of data from many sources, including books, journal or conference articles, theses or dissertations, and the Internet. While these data streams can be beneficial to an individual, they can become too heavy and cumbersome for a single person. This phenomenon can eventually become discouraging to some. Information overload has been identified as a major factor contributing to the poor academic performance of many students.

To address the problem of information overload, computer science is investigating ways in which computers or algorithms can help humans manage information overload. In particular, this field has developed information search and filtering techniques that allow information to be tailored to individual tastes, needs, and preferences. More recently, recommender systems have been employed

© Springer Nature Switzerland AG 2022
M. Fakir et al. (Eds.): CBI 2022, LNBIP 449, pp. 107–121, 2022.
https://doi.org/10.1007/978-3-031-06458-6_9

to offer only information that is likely to be of interest to their users, omitting those that do not fit their profile.

We believe that recommender systems can be very beneficial to students searching for books in their university library. Indeed, this environment contains a huge amount of resources, books, periodicals, scientific journals, theses and dissertations.

Although there are many types of recommender systems, the first step in developing one is to determine what types of data will be used. An explicit information [1] system will use data where users have indicated their interests in products. There are several ways for a user to express his preferences for a product. He can do this by using a vote, which is a scale with a predefined scale and is often used for online product sales. Also, on some websites, it is possible to register in order to create a profile and choose from a list of preferences, those that are closer to our personal preferences. In some cases, in addition to creating a profile, it is possible to create a wish list. The purpose of this list is to indicate products that the user is interested in but has not yet purchased.

On another note, an implicit information system will use various sources of information without having to ask the user anything. A popular example of implicit information is transaction history [20]. Other information such as the time a user spends on a product's description page, the history of products visited, and the time spent on certain pages can all be used to make the system more efficient.

There are several types of approaches to implementing a recommender system. The approaches presented in this section apply to both explicit and implicit The approaches presented in this section apply to both explicit and implicit recommender systems. Formally, a recommender system is a data filtering tool where there is a set of users U a set of products P and a utility function h which denotes the interest of a user for a product. Thus, it is possible to see recommendation systems as a function to be maximized:

$$\forall u \in U, i'_u = argmax_{x \in I} h(u, i)$$

I represents a set of products for U and the utility function h is particular to the type of approach that is used. Also, this function varies according to the type of information used, either implicit or explicit. The objective is thus to estimate the utility function of a document i for a user u.

2 Related Previous Works

Neural networks have brought important advances to several application domains and it is not surprising that the recommender systems community has also been looking at them. Community has also turned its attention to them.

$$\hat{R}_{ui} = \theta_u^T \beta_i$$

The objective function is a combination of the difference between the approximation and the objective function is a combination of the difference between the approximation and the actual value, and an L2 regularization of the latent factors for users and items.

$$L_{MF} = \sum_{u,i} c_{ui}(R_{ui} - \theta u^T \beta_i)^2 + \lambda_\theta \sum u \|\theta u\|\, 2^2 + \lambda\beta \sum i \|\beta i\|\, 2^2$$

Wang et al. (2015) [9] propose one of the first methods combining recommender systems with neural networks. They present a Bayesian hierarchical model consisting of two parts. The first creates a vector representation of the item using a Stacked Denoising Autoencoder (SDAE), a neural network encoding a corrupted version of the input data into a smaller representation and then using this as input to a decoder reconstructing the original data. The latent representation given by the decoder becomes the item representation. The model is evaluated on two article citation datasets and a third from a contest presented by Netflix. The comparison is made with other hybrid recommendation systems and shows that the author model performs consistently better than the others.

Liang et al. (2016) [6] builds on the success of models using vector-based word representations (word2vec, GloVe, etc.) to improve a vector representations of words (word2vec, GloVe, etc.) to improve a recommendation system using matrix factorization with collaborative filters. This factorization de- the score matrix into a product of latent factors for users and items. The objective function is a combination of the difference between the approximation and the true value, and a and an L2 regularization of the latent factors for users and items. In addition, the SPPMI (shifted positive pointwise mutual information) matrix is computed, representing the co-occurrence patterns between each item and its context. Representing the co-occurrence patterns between each item and its context.

Gong and Zhang (2016) [8] use convolutional neural networks (CNNs) with attention to recommend hashtags using the text of a tweet as input. Attention is used to only give importance to the most informative words. Zheng et al. (2017) [14] uses two CNNs to encode user behavior and item attributes based on the text of reviews before combining them through a factorization machine at the last layer. Hidasi et al. (2015) uses a recurrent neural network (RNN) to recommend a user's next action using their previous actions. Hidasi et al. (2016) [3] extend this model by using additional information to create RNNs each encoding a type of data before concatenating the results to predict the next item. Smirnova and Vasile (2017) [7] also use additional information, but use it directly as input to an RNN predicting next clicks. Wu et al. (2017) [4] use an RNN for the score given by a user to an item, taking into account that the attractiveness of an item and a user's preferences change over time.

Liu et al. (2018) [21] use explicit user and item information to predict whether a user has rated an item. Their model is separated into two stages. In the first stage, an autoencoder attempts to reconstruct a user's attributes and the intermediate layer is concatenated to an embedding of the user, i.e. a vector representing him/her, whose elements are elements are obtained during training. The same thing is done for the item with an autoencoder and a different embedding. Subsequently, two different methods are used to predict the output. The first one, GMF++, multiplies these two vectors by elements with to pass the result in a multilayer perceptron. The second, MLP++, concatenates them rather than multiplying them. The loss function is a sum of the loss function of each of the two each of the two autoencoders and the cross-entropy loss between the output of the model and the actual existence or not of an item evaluation by the user.

3 Materials and Methods

3.1 Collaborative Filtering

There are different approaches to recommender systems. They are based on factors such as user profile, knowledge of the resources to be recommended, similarities between profiles of different users, etc. The approaches can be classified into three main categories: content-based recommendations, collaborative filtering and hybrid approaches [1,11,17].

In this study, we will focus on recommendation based on collaborative filtering. We will study a very used method, the matrix factorization (decomposition in two matrices of latent factors) based on neural networks.

Collaborative filtering [5] is a technique that uses the collaboration of elements (such as users) to perform a task, such as issuing a recommendation. Collaborative filtering works as follows. First, the system collects information (such as ratings or purchase history) about the users in order to obtain their preferences. Then, the system will compare the collected information with other users and find the most similar ones. When the similar users are found, the system will recommend products (that the similar users have voted for, but that the targeted user has not purchased) to the user. The main assumption behind collaborative filtering is that if a user likes a product, then users with similar tastes to that user will also like that product. For example, if user A shares the same preferences as user B and A buys a book on Amazon and gives it a good vote, then this book will be recommended to B.

One of the most popular methods of collaborative filtering is the k nearest neighbor (KNN) algorithm [15]. KNN is a classification algorithm where k is a parameter that specifies the number of neighbors used. To use KNN, the system must have a similarity measure to distinguish between users who are close and those who are far away. Depending on the problem studied, this similarity measure can be derived from the Euclidean distance, the cosine measure, the Pearson correlation, etc. Then, when the system is queried about a new piece of information, for example a search for products that might appeal to an individual, it will go and find the k users who are closest to the target user. Finally, a majority vote is taken to decide which product will be recommended. For example, in the case of an explicit information system, the system could recommend the product with the highest rating. While for implicit information, the system could recommend the most popular product in the neighborhood.

KNN is a fairly simple algorithm to understand. Mainly because it does not need a model to make a prediction. The counter cost is that it must keep in memory all the observations to be able to make its prediction. So we have to be careful with the size of the training set. Also, the choice of the method to compute the distance and the number of neighbors may not be obvious. It is necessary to try several combinations and to do some training of the algorithm to have a satisfactory result.

3.2 Latent Model

Singular Value Decomposition (SVD) and Principal Component Analysis (PCA) [22] are well established techniques for identifying latent factors in the field of information retrieval to face the challenges of collaborative filtering (CF). The principle of Matrix Factoring is one of the most widely used algorithms in collaborative filtering that belongs to the model-based methods [10].

Any matrix $R \in \mathbb{R}^{u \times i}$ can be decomposed as follows:

$$R = U\Sigma V^T$$

with $U \in \mathbb{R}^{u \times u}$ the eigenvectors and orthonormals of RR^T, $V \in \mathbb{R}^{i \times i}$ the eigenvectors and orthonormal eigenvectors of $R^T R$, and $Sigma \in \mathbb{R}^{u \times i}$ the diagonal matrix containing the square roots of the eigenvalues of R.

These matrices can be used in two main ways in the context of recommender systems: by approximating the R matrix or by creating a new vector basis of latent factors. The R matrix can be approximated by:

$$R' = U_d \Sigma_d V_d^T$$

where U_d, V_d and Σ_d represent respectively the first d eigenvectors of the columns, eigenvectors of the rows and eigenvalues. R' is a dense matrix where each element can be taken as a prediction of the corresponding of the item of the corresponding column for the user of the corresponding row.

The other use of the SVD decomposition is to create a new vector basis of latent factors of columns (or rows) obtained by $U_d \Sigma_d$ (or $\Sigma_d V_d^T$). This reduction allows to have a potentially much smaller dense matrix that can be used in future similarity similarity calculations. This reduction in latent factors also potentially allows a better generalization of the data since users or items but not necessarily identical in the original vector database will also be similar in the in the new vector base of latent factors. A user $x \in \mathbb{R}^i$ (or item $y \in \mathbb{R}^u$) can be transformed into the new basis by $x' = V_d^T x$ (or $y' = U_d^T y$).

For example, in a movie recommendation context, these latent factors obtained will potentially represent the genre of the movies and allow to see that two users are listening to the same type of users are listening to the same type of movies [16], without requiring that the same particular movies have been have been listened to by these two users [11].

3.2.1 The SVD Algorithm

The famous SVD algorithm, popularized by Simon Funk during the Netflix Prize. When baselines are not used, this is equivalent to probabilistic matrix factorization[1,2]. The prediction $\hat{r}_{ui}$ is defined as:

$$\hat{r}_{ui} = \mu + b_u + b_i + q_i^T p_u$$

---

[1] https://github.com/NicolasHug/Surprise/.

[2] https://surprise.readthedocs.io/en/stable/index.html.

If the user u is unknown, then the bias bu and the factors p_u are assumed to be zero. The same applies for the item with b_i and q_i.

To estimate the entire unknown, we minimize the following regularized squared error

$$\sum_{r_{ui} \in R_{train}} (r_{ui} - \hat{r}_{ui})^2 + \lambda \left(b_i^2 + b_u^2 + ||q_i||^2 + ||p_u||^2\right)$$

To minimize the function L, we will use two algorithms: Alternating Least Square (ALS) and Stochastic Gradient Descent (SGD).

3.2.2 Alternate Least Squares

The alternating least squares[3] method consists in fixing one of the latent factors [10] (X or Y) and computing the gradient with respect to the other factor. By cancelling the gradient, we find the rule for updating one of the factors. By doing the same operation again, we can easily determine the other factor. Let us try to derive the objective function. Let's fix y_i and cancel the derivative:

$$\frac{\partial \mathcal{L}}{\partial x_u} = -2 \sum_i (r_{ui} - x_u^T \cdot y_i) y_i^T + 2\lambda x_u^T$$

$$\lambda x_u^T - (r_u - x_u^T Y^T) Y = 0$$

$$r_u Y = x_u^T (Y^T Y + \lambda I)$$

We deduce that:

$$x_u^T = r_u Y (Y^T Y + \lambda I)^{-1}$$

Fixing x_i, the same reasoning allows us to write:

$$y_i^T = r_i X (X^T X + \lambda I)^{-1}$$

For each iteration of the algorithm, we fix one of the factors and calculate the other. We define a rule for updating the factors by repeating these iterations until there is convergence.

The minimization of the objective function by stochastic gradient descent was popularized by Simon Funk. As in the previous algorithm, we first initialize the latent factor matrices X and Y. We define a training set on which we review all the evaluations of the starting matrix. At each iteration, we evaluate the prediction error e_{ui} defined as follows:

$$e_{ui} = r_{ui} - \hat{r}_{ui}$$

We update the latent variables x_u and y_i in the opposite direction to the gradient with a step *gamma*.

$$x_u \longleftarrow x_u + \gamma(e_{ui}y_i - \lambda x_u)$$

$$y_i \longleftarrow y_i + \gamma(e_{ui}x_u - \lambda y_i)$$

---

[3] https://github.com/microsoft/recommenders.

Biases in the recommendation[4] Some books tend to receive higher ratings than others because of their popularity. We also see that some users are more generous in rating books than others. Let's take the example of two users A and B. Let's say that A's average rating is 1.5 and B's is 4.7. A rating of 3 for a new book by user A does not have the same meaning as a 3 by user B. A probably enjoyed the new book while B did not. This is why we introduce biases that we associate with users and books. Taking these biases into account considerably increases the relevance of the recommended items.

The bias b_{ui} associated with the score r_{ui} is defined by: $b_{ui} = \mu + b_u + b_i$

- μ is the average of the evaluations
- b_u user bias
- b_i bias associated with films

From then on, the prediction of the score of the item i by the user u is rewritten as

$$\hat{r}_{ui} = x_u^T y_i + b_u + b_i + \mu$$

The objective function to be minimized is therefore:

$$\mathcal{L} = \sum\nolimits_{(u,i) \in \Gamma} (r_{ui} - x_u^T \cdot y_i - b_u - b_i - \mu)^2 + \lambda(||x_u||^2 + ||y_i||^2 + b_u^2 + b_i^2)$$

3.3   Neural Network Model

3.3.1   Multilayer Perceptron

The multilayer perceptron [13] is the basic element of most neural networks [18]. To describe neural networks, we will begin by describing the simplest possible neural network, one which comprises a single "neuron" (Fig. 1).

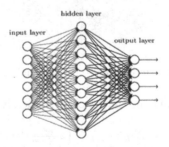

Fig. 1. Diagram to denote a single neuron

As its name indicates, it is composed of several layers [12] all following the same pattern, i.e. a linear transformation followed by a function applied per element, is called the activation function. More concretely, for an input data

---

[4] https://surprise.readthedocs.io/en/stable/matrix_factorization.html#surprise. prediction_algorithms.matrix_factorization.SVD.

vector $x \in R^p$ a weight matrix $W \in R^{k \times p}$, a bias vector $b \in \mathbb{R}^k$ and an output $\hat{y} \in \mathbb{R}^k$, we have

$$z = Wx + b$$
$$y = \sigma(z)$$

where *sigma* represents any function applied per element. Some of the most commonly used are (Fig. 2):

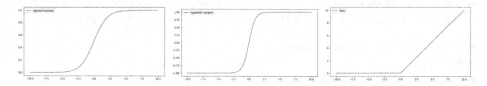

Fig. 2. The main activation functions that can be found in the field of neural networks

sigmoid: $f(x) = \frac{1}{1+e^{-x}}$, tanh : $f(x) = \frac{e^x - e^{-x}}{e^x + e^{-x}}$, relu: $f(x) = \max(0, x)$

When there is more than one layer, the output of one is the input of the next. Let $g(x)$ be the result of the equation $y = \sigma(z)$. The result *haty* will be given by:

$$h^{(0)} = g(x)$$
$$h^{(1)} = g(h^{(l-1)})$$
$$\hat{y} = h^{(L)}$$

where L is the number of layers (Fig. 3).

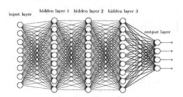

Fig. 3. We have used circles to also denote the inputs to the network

For a classification problem, the function typically used at the last layer is softmax:

$$softmax(y) = \left(\frac{e^{y_i}}{\sum_k e^{y_k}} \right)_i$$

In order to adjust the weight and bias parameters, the model must be trained, i.e. given correct input and output data pairs and made to make the best possible predictions from them.

3.3.2   Backpropagation Algorithm

The typical way to train neural network models is through gradient descent and backpropagation. To do this, we need to define a cost function and compute the gradient of each parameter with respect to it.

Two widely used cost functions are the quadratic distance for regression problems and regression problems and the cross entropy for classification problems. For a regression problem and outputs and predictions given by $y_i, \hat{y}_i \in R$, the distance quadratic distance is given by:

$$J = \sum_i (\hat{y}_i - y_i)^2$$

Similarly, for a classification problem and outputs and predictions given by $y_i, \hat{y}_i \in \mathbb{R}^k$, the cross entropy is

$$J = -\sum_i \sum_j y_{ij} \log \hat{y}_{ij}$$

Backpropagation consists in defining the gradients of each layer with respect to the upper layers and thus being able to propagate these through the network. An example of this can be seen in Fig. 4. Consider a regression problem with the quadratic distance given by equation $J = \sum_i (\hat{y}_i - y_i)^2$ as the cost function. With input data x and output data y, we will define the different layers of the MLP by:

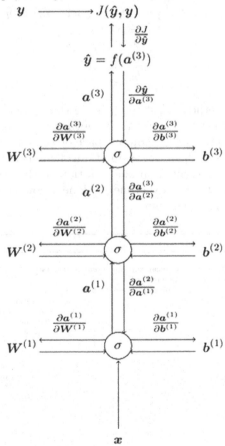

Fig. 4. Example of backpropagation

We can see that the output of each layer is defined with respect to the lower layer and the gradient is defined with respect to the upper layer. At each layer, σ represents the operation $a^{(i+1)} = g(a^{(i)}W^{(i+1)} + b^{(i+1)})$ with g a function applied per element and f a problem-dependent function (e.g. softmax for a classification problem).

$$z_1 = W_1^T x + b_1$$
$$a_1 = \sigma(z_1)$$
$$z_1 = W_l^T a_{l-1} + b_l$$
$$a_1 = \sigma(z_1)$$
$$\hat{y} = w_L^T a_{L-1} + b_L$$

where σ is the sigmoid function defined in equation $f(x) = \frac{1}{1+e^{-x}}$. The gradients will be:

$$\frac{\partial J}{\partial \hat{y}} = 2(\hat{y} - y) \equiv \delta_L$$
$$\frac{\partial J}{\partial z_{L-1}} = (\delta L w_L) \odot aL - 1 \odot (1 - a_{L-1}) \equiv \delta_{L-1}$$
$$\frac{\partial J}{\partial z_1} = (W_{l+1}^T \delta_{l+1}) \odot a_1 \odot (1 - a_l) \equiv \delta_l$$
$$\frac{\partial J}{\partial w_L} = \delta L aL - 1$$
$$\frac{\partial J}{\partial b_L} = \delta_L$$
$$\frac{\partial J}{\partial W_l} = a_{l-1} \delta_L^T$$
$$\frac{\partial J}{\partial b_l} = \delta_l$$

Although several methods exist to optimize the parameters using these gradients, recent approaches use a stochastic optimization, i.e. using only a part of the data at each iteration to compute part of the data at each iteration to compute the gradients and an approximation of moments to ensure that the gradients are not too different from one iteration to the other. A very popular algorithm that is used for the experiments done below is Adam (Kingma and Ba, 2014)[5], described in Algorithm 1.

Algorithm 1. Algorithm Adam

Require: $\alpha > 0$: Step size
Require: $\beta_1, \beta_2 \in [0, 1)$: Exponential decay rate for moment estimates
Require: $\epsilon > 0$: Digital stability parameter
Require: $L(\theta)$: Objective function to be minimized with parameters θ
Require: θ_0 :   Initial parameters θ
$m_0 \leftarrow 0$ Initialization of the vector of 1st moments
$v_0 \leftarrow$ Initialization of the 2nd moment vector
$t \leftarrow 0$ Initialization of the time interval
While θ_t did not converge **do**
$t \leftarrow t + 1$
$g_t \leftarrow \nabla_\theta L_t \theta_{t-1}$ Obtaining the gradients
$m_t \leftarrow \beta 1 \cdot mt - 1 + (1 - \beta_1) \cdot g_t$ Updating
$v_t \leftarrow \beta 2 \cdot vt - 1 + (1 - \beta_2) \cdot g_t^2$ Updating
$\hat{m}_t \leftarrow \frac{m_t}{1 - \beta_1^t}$ Updating
$\hat{v}_t \leftarrow \frac{v_t}{1 - \beta_2^t}$
$\theta_t \leftarrow \theta_{t-1} - \alpha \cdot \frac{\hat{m}_t}{\sqrt{\hat{v}_t} + \epsilon}$
end While
return θ_t

---

[5] https://arxiv.org/abs/1412.6980.

This one uses estimates of the first two moments of the gradients to have more constant updates of the parameters from one iteration to the next.

4   Experiments

4.1   Dataset

The data is preprocessed accordingly in the file via the Pandas module. The simple recommender has been extended to load and store the goodbooks-10k dataset[6], in addition to BX-Book-Crossings. The implemented files for the dataset repair already provide the train-, validation- and test split accordingly and is imported into all necessary files. Our database is composed of 53424 users and 10000 books. We see that there are almost 6 million ratings and the Average rating is 3.92 with the Median rating of 4 which tells me that the overall books are rated fairly highly and the data is likely evenly distributed without significant variations. The ratings are on a scale of 1 to 5.

4.2   Training

To train the model, we need to define a loss function to be optimized. The training is done in an iterative way by taking at each iteration a sample of size b of the not null values of this collaborative matrix C.

This will give a vector of values $c_i, i = 1, \ldots, b$ Then vectors u_i and v_i are computed by the above model for the users corresponding to the row and column of the element ci in the matrix C.

The loss function is composed of two parts. The first is a measure of the distance between these two vectors ui and vi weighed by the desired approximation ci. The greater the closeness is important, the more this distance will have an important weight for the loss.

4.3   Evaluation Metrics

Precision and recall are the two most popular measures for evaluating an information retrieval system and the F-Measure strikes a balance between precision and recall (Table 1).

Model	Epoch @	Loss	Learning rate	The regression loss	Accuracy of the model
MFM	@100	0.5904	0,009		
	@200	0.3792	0,009	1.6648	12.08%
Biased MatFac	@100	0.6764	0,009		
	@200	0.3792	0,009	1.2351	22.25%
NeuralNetwork	@100	0.5930	0,009		
	@200	0.3665	0,009	0.9159	21.84%

---

[6] https://www.kaggle.com/zygmunt/goodbooks-10k.

It is worth noting that both Matrix Factorization models have very similar distribution of embedding weights and that some outliers can be clearly spotted, while the bulk of the weights clearly centers around a certain space. [2].

About neural networks, Interestingly enough, the outliers stil exist, but with more weights apparent, they seem to be stil as much as in the other two models, while a good chunk of the weights is in the middle (Fig. 5, Table 1).

Table 1. Precision, Recall, F1-score and Support comparison between MFM, Biased MatFac and Neural network algorithms. The training set consists of 80% of the data in the goodbooks10k dataset, while validation and training make up 10% each.

	MFM				Biased MatFac				Neural network			
Precision	0.02	0.06	0.25	0.35	0.02	0.06	0.25	0.37	0.00	0.07	0.25	0.37
Recall	0.02	0.66	0.32	0.00	0.01	0.22	0.70	0.08	0.00	0.02	0.31	0.52
F1-score	0.02	0.12	0.28	0.01	0.01	0.10	0.37	0.14	0.00	0.03	0.31	0.43
Support	1792	6259	24420	36049	1792	6259	24420	36049	1792	6259	24420	36049

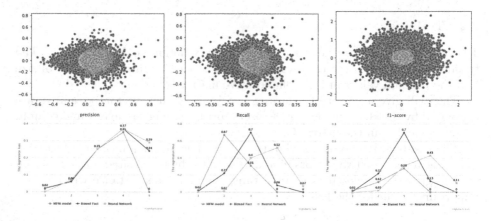

Fig. 5. Extracting embeddings and embedding weights MFM model, BiasedMATFAC and NeuralNetwork

5   Results

The results of our experiment clearly show that the MFM model has difficulty in fitting, unlike the model based it is almost close to the results of the neural network model. According to the graphs of recall and accuracy then the F-score the neural network model is more powerful than the two other model, our training was on a desk core i5 11 generation 16go of Ram. An improvement of our model in the future will give us more accurate results to say that really the concept of the nursey trainers is more powerful.

We will compare our approach with models reported in a journal published in 2020 that uses Recurrent Neural Networks (RNN) [19], and we want to work on this for an extended version of our paper.

Model	Accuracy	Recall @10
course2vec	0.5011	0.7685
ins-course2vec	**0.5138**	0.7853
dept-course2vec	0.3581	0.8284
ins-dept-course2vec	0.4961	**0.8557**
tf	0.3037	0.537
binary	0.3159	**0.581**
tf-idf	**0.3227**	0.542
tf+insdept-course2vec	0.5066	0.8438
tf+insdept-course2vec (norm)	0.448	0.6872
bin.+insdept-course2vec	**0.5193**	**0.8788**
bin.+insdept-course2vec (norm	0.4603	0.7449
tfidf+insdept-course2vec	0.5138	0.8584
tfidf+insdept-course2vec (norm)	0.4503	0.7059

The multifactor model outperformed the pure race2vec model in terms of recall@10 in both validation sets. The combined instructor and department factor model gave the best. The same result was obtained for the mean and median rank measures in equivalence, but the multifactor models were not always the best in terms of analog recall. The best in terms of analog recall@10 (Accuracy).

6 Conclusion

Collaborative filtering with Neural Networks and Matrix Factorization, on the other hand, provides for an interesting method implementing and realizing the task of collaborative filtering as it primarily works with embeddings. However, my findings showed that Neural Collaborative Filtering depends on a strong database where as much of observations of each rating has to be given, since otherwise the predictions will be inaccurate to the model being biased towards ratings that is sees more often. Another problem is Bias/Variance since greatly overfitting the model will result in inaccurate predictions which thus negatively affects the final recommendations. The Neural Network model seemed to have the least problems with this, however, it also could have performed better, and given approprite conditions (e.g. strong CPU which can quickly calculate 1000 epochs), it could also provide better results. Neural Networks provide for a great opportunity in experimentally handling the task of user recommendations. One last thing to note is that all three models suffer from the very nature of Collaborative filtering, in that they tend to recommend the most popular items on a given platform. While this fulfills the task of Collaborative filtering, it actually might not fulfill the expectations of some users whose tastes go against the mainstream.

References

1. Aggarwal, C.C., et al.: Recommender Systems. Springer, Cham (2016). https://doi.org/10.1007/978-3-319-29659-3
2. Anelli, V.W., Bellogín, A., Di Noia, T., Pomo, C.: Reenvisioning the comparison between neural collaborative filtering and matrix factorization. arXiv:2107.13472v1. 28 July 2021
3. Baltrunas L., Tikk, D., Hidasi, B., Karatzoglou, A.: Session-based recommendations with recurrent neural networks (2015)
4. Beutel, A., Smola, A.J., Jing, H., Wu, C.-Y., Ahmed, A.: Recurrent recommender networks, pp. 495–503 (2017)
5. Chen, J., Zhang, H., He, X., Nie, L., Liu, W., Chua, T.-S.: Attentive collaborative filtering: multimedia recommendation with item-and component-level attention. In: DANS Proceedings of the 40th International ACM SIGIR Conference on Research and Development in Information Retrieval, pp. 335–344 (2017)
6. Charlin, L., Blei, D.M., Liang, D., Altosaar, J.: Factorization meets the item embedding: regularizing matrix factorization with item co-occurrence, pp. 59–66 (2016)
7. Smirnova, E., Vasile, F.: Contextual sequence modeling for recommendation with recurrent neural networks. In: Proceedings of the 2nd Workshop on Deep Learning for Recommender Systems, pp. 2–9 (2017)
8. Gong, Y., Zhang, Q.: Hashtag recommendation using attention-based convolutional neural network. In: IJCAI, pp. 2782–2788 (2016)
9. Yeung, D.-Y., Wang, H., Wang, N.: Collaborative deep learning for recommender systems. In: DANS Proceedings of the 21th ACM SIGKDD International Conference on Knowledge Discovery and Data Mining, pp. 1235–1244. ACM (2015)
10. Wang, N., Wang, H., Yeung, D.-Y.: KDD. Collaborative deep learning for recommender systems, vol. 22(1) (2015)
11. Wang, X.S.H., Yeung, D.-Y.: Collaborative recurrent autoencoder: recommend while learning to fill in the blanks. In: NIPS (2016)
12. He, X., Liao, L., Zhang, H.: Neural collaborative filtering. arXiv:1708.05031v2, 26 August 2017
13. He, X., Liao, L., Zhang, H., Nie, L., Hu, X., Chua, T.-S.: Neural collaborative filtering. In: WWW, pp. 173–182 (2017)
14. Yu, P.S., Zheng, L., Noroozi, V.: Joint deep modeling of users and items using reviews for recommendation, pp. 425–434. ACM (2017)
15. Lemire, D., Maclachlan, A.: Slope one predictors for online rating-based collaborative filtering, 24 February 2007
16. Ning, X., Karypis, G.: SLIM: sparse linear methods for top-n recommender systems. In: ICDM, pp. 497–506 (2011)
17. Trofimov, M.: Representation learning and pairwise ranking for implicit and explicit feedback in recommendation systems (2017)
18. Wang, H., Wang, N., Yeung, D.-Y.: Collaborative deep learning for recommender systems. In: DANS Proceedings of the 21th ACM SIGKDD International Conference on Knowledge Discovery and Data Mining, pp. 1235–1244 (2015)

19. Zachary, A., Pardos, W.J.: Designing for serendipityina university course recommendation system (2020)
20. Zhang, H., Nie, L., Hu, X., He, X., Liao, L., Chua, T.-S.: Neural collaborative filtering. https://github.com/maciejkula/spotlight
21. Khan, M.S., He, J., Liu, Y., Wang, S.: A novel deep hybrid recommender system based on auto-encoder with neural collaborative filtering (2018)
22. Zhang, S., Yao, L., Sun, A.: Deep learning based recommender system: a survey and new perspectives. arxiv preprint arxiv:1707.07435 (2017)

Comparative Study of Deep Learning Models for Detection and Classification of Intracranial Hemorrhage

Lale El Mouna[1](✉), Hassan Silkan[1], Youssef Haynf[2], Amal Tmiri[3], and Abdellatif Dahmouni[1]

[1] Laroseri Laboratory, Chouaib Doukkali University, El Jadida, Morocco
laleelmouna@gmail.com
[2] The National School of Business and Management of Dakhla, Ibn Zohr University, Agadir, Morocco
[3] Ensam, Rabat, Morocco

Abstract. Today's technologies have deeply influenced human health and daily life. Consequently, the health care process is widely improved to automate and detect diseases. Deep Learning and Transfer Learning classifiers are within emergent technologies that impact health care. In this paper, we used Transfer Learning and Convolutional Neural Network to classify and detect the Intracranial hemorrhage (ICH). The performance of the used classifiers is evaluated and compared on the Intracranial Hemorrhage Dataset that contains 2814 images. The results show that the detection accuracy of Transfer Learning with Inception V3, which achieves 88.97%, is superior to that of the Convolutional neural network.

Keywords: Deep learning · Classification · Convolutional neural networks

1 Introduction

Intracranial hemorrhage (ICH) is one of the important field of search in medical imaging. It is a dangerous disease that requires a rapid diagnosis to speed up treatment and save the patient's life. ICH is classified into five types: epidural (EDH), subdural (SDH), intraventricular (IVH), subarachnoid (SAH) and intraparenchymal (IPH) (Fig. 1).

The use of deep learning is developing more and more and is becoming more and more important in many sectors of activity. In Intracranial hemorrhage (ICH), the appearance of this new technology appears to be a real revolution. The uses of this technology are numerous and should, in the long term become essential in this field of search.

In this work, we propose to classify and detect the Intracranial hemorrhage (ICH) by using two convolutional neural network methods of deep learning techniques CNN and transfer learning. We have used an ICH database composed of 2814 images and we have augmented Database by generate more images by applying some geometric transformation such as rotation, translation, transvection (shear mapping) and horizontal flipping.

© Springer Nature Switzerland AG 2022
M. Fakir et al. (Eds.): CBI 2022, LNBIP 449, pp. 122–131, 2022.
https://doi.org/10.1007/978-3-031-06458-6_10

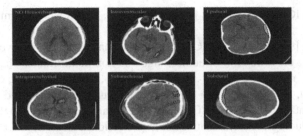

Fig. 1. Different types of intracranial hemorrhage [1]

Our work is organized as following. In Sect. 2, we presents some related works. In Sect. 3, we describe the methodology used in this paper. A results of the different methods are reported in Sect. 4. The last section discusses the obtained results and we finish up with a conclusion.

2 Related Works

Detection and classification of intracranial hemorrhage (ICH) is one of the important research areas. In the literature, many methods based on deep learning have been proposed. CNNs have won popularity due to their reliance and effectiveness and have become important constraints in medical diagnosis [1–3]. The proposed methods for ICH detection can be divided into three principal categories: single class detection relating to a single ICH class, multi-class classification distinguishing ICH subtypes, and ICH pixel area detection in individual images [1, 4, 5]. Based on hemorrhage classification, these studies can be divided into five types: epidural, subdural, subarachnoid, intraparenchymal and intraventricular.

The proposed method in [1], consist to detect the intracranial hemorrhage and identify its subtypes by combining 3D CNN and recurrent neural network. The accuracy obtained show that its values varied between 75% and 96% depending on the subtype and the identification of any hemorrhage exceeded 98%.

In [8], Ker et al. proposed a 3D CNN to classify the five subtypes of ICH: healthy brain, SAH, IPH, brain polytrauma hemorrhage (BPH) and subdural hemorrhage (SDH). The Dataset used, contains approximately 12000 images, is comprised 399 computed tomography scans. The measure of performance of Two sorting classifications are considered. The first one is a result of two classes (normal with specific ICH), while the second performance is obtained by considering four classes. The obtained results by using F1 score show that the performance of two classes varies between 70.6% and 95.2% while the performance in multiclass analysis is 68.4%

In [6], the authors propose a combination of CNN, autoencoder network and a heatmap method. The dataset used contains 2101 images with augmentation. The heat maps were created for each image. The autoencoder network was used to process the data and a support vector machine was used to classify the data. The obtained results show prove an accuracy of 98.6%, a sensitivity of 98.1%, and a specificity of 99.0%.

Burduja et al. [5] proposed a classify ICH subtypes method by combining ResNeXt-101and bidirectional long short-term memory. The obtained on test dataset prove that the accuracy in ICH subtypes was greater than 96%, with a weighted mean log loss of 0.04989.

Castro et al. [9] employed 491 CT scans to train and assess two convolutional neural networks for the classification of hemorrhage and non-hemorrhage. The planned CNN networks have a recall rate of 97%, accuracy of 98%, and an F1 metric of 98%.

Lewiki et al. [10] developed a convolutional neural network based on ResNet the dataset used contains 752,803 DICOM files. The obtained accuracy is a 93.3% accuracy in case of producing the correct multiclass prediction and a 76% in case of average per-class recall score.

In [11], the authors present a mixture of a convolutional neural network and a long short-term memory. the long short-term memory connected features across slices, while the pre-trained Convolutional neural networks extracted features from each image.

Chilamkurthy et al. [12] proposed four techniques for detecting ICH subtypes. They used a huge dataset containing 290k and 21k CT scans to train and evaluate their algorithms. For the testing, two datasets were employed. One of them had 491 scans and was released to the public (called CQ500).

3   Methodology

In this paper we proposed to treat the problem of intracranial hemorrhage by comparing the Deep learning technics, in our work we have realized a comparative study by choosing two methods of CNN, in next section we present the architecture of each one.

To carry out this project, we divided it into two parts. The first part is to detect the hemorrhage whereas the second part is to set up models that would allow us to classify the different types of hemorrhage. In this part, we will explain the methods used to achieve each of the two tasks.

3.1   Convolutional Neural Networks

Convolutional neural networks are the most common type of neural network, which are used to analyze high-dimensional data such as images and videos [13].

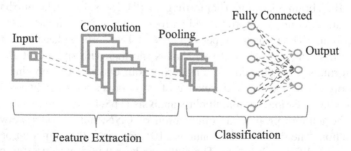

Fig. 2. Basic architecture of CNN

It's a multi-layer neural network (NN) inspired on visual cortex neurology, with a convolutional layer and a fully connected (FC) layer (s). Pooling layers can occur between these two levels (see Fig. 2). They outperform Deep Neural Networks (DNNs), which challenge to scale well when dealing with multidimensional, locally linked input data. As a result, the most common application of CNN is in datasets with a large number of neurons and parameters to learn (e.g., image processing).

CNN Model

In a first step, we tested a usual convolutional neural network, which is made of five layers. Each layer of contains a convolution and a Maximum pooling. Afterwards, we connected the output of layers with a function that allows to flatten the result, called flattening. Then comes a Dense neural network, that is, a network that contains several layers completely connected. For the latter, we put two hidden layers with a Sigmoid activation function for the last layer in order to determine the probability of belonging to each class.

3.2 Transfer Learning

The amount of dataset used in medical image analysis and processing is one of the major problems. Therefore, using a small quantity of data to train a fully deep network structure like CNN can lead to overfitting [14].

To solve the basic problem of insufficient training data, we use Transfer Learning, its principle consist to transfer the knowledge from the source domain to the target domain by relaxing the assumption that the training data and the test data must be independent and identically distributed. Indeed, Transfer learning use a pre-trained model and transferring its parameters to the targeted network model. Afterwards, we connected the last fully layers trained with initial random weights on the new dataset, in order that the model could be used for another task. Despite the fact that the dataset is not the same as the one used to train the network, the low-level features are the same.

Therefore, transferring the parameters, specifically weights, of the pre-trained model to targeted network model could provide it with a powerful feature extraction.

The obtained results in medical imaging, prove the performance of transfer learning model and its superiority vis-à-vis several methods proposed in the literature [15–17]. The pre-trained model used to classify intracranial hemorrhage is presented in the following section (inception v3).

Inception V3

Inception-v3 model is an advance convolutional neural network invented by Szegedy et al. [18] from the Google [19]. It used transfer learning to produce good classification performance in a variety of medical applications [20, 21]. Inception-v3 included an Inception model that, like Goog-LeNet, blends many convolutional filters of various sizes into a single filter. This architecture decreases the number of parameters that must be trained and, as a result, reduces computational complexity. Figure 3 presents the Inception v3 architecture in its basic form.

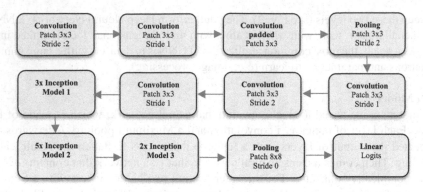

Fig. 3. The basic architecture of Inception-v3 [22].

3.3 Data Process

Dataset

The used database contains 82 computerized Tomography (CT) images of 36 scans for patients diagnosed with ICH of five types: Subdural, Epidural, Subarachnoid, Intraventricular and Intraparenchymal. Each computerized Tomography scan has around 30 slices with a thickness of 5 mm. Patients 'ages ranged from 27.8 to 19.5, with a mean of 27.8 and a std of 19.5. Males made up 46 of the patients, while females made up 36. Two radiologists examined each slice of a non-contrast CT scans, identifying various types of hemorrhage and fractures that occurred. Each slice's ICH areas were likewise marked by the radiologists. The radiologists had reached an understanding. radiologists did not have access to the patients' clinical histories.

The reading and using database are done in two steps. Firstly, the Syngo by Siemens Medical Solutions [23], was utilized to read the computerized Tomography DICOM files, save the content of a brain window with level = 40 and width = 20, and save the content of bone window with level = 700 and width = 3200. In Second step, a set of tools was developed for read the avi files and perform the annotation (Table 1).

Table 1. The number of slices of Intracranial hemorrhage sub-types

ICH sub-types	Slices
Intraventricular	24
Intraparenchymal	73
Subarachnoid	18
Epidural	182
Subdural	56
No hemorrhage	2173

Data Augmentation

We employed data augmentation, a typical deep learning technique, to partially alleviate the need to supply the network with a vast amount of data. This method enables for the presentation of image in multiple aspects.

We used a geometric augmentation by applying for each image a set of geometrical transformation such as translation, transvection (shear mapping) and horizontal flipping whose parameters are random. The goal is to developed a model with variations of the information accessible in the original images at each iteration.

4 Results

The two methods have proven their effectiveness by their results in the two tasks (detection and classification), the two models are launched on a data which contains 2814 images, divided into two sets. Training dataset containing 80% of dataset and test database containing 20% of dataset.

4.1 Evaluation Metric

In order to verify the performance of the proposed method, the accuracy were calculated. It is the measure of how accurately the classifier can classify the data. its formula is given as follows:

$$Accuracy = (TP + TN)/(FP + FN + TP + TN)$$

where
TP: correctly predicted positive class
FP: incorrectly predicted positive class
FN: incorrectly predicted negative class
TN: correctly predicted negative class

4.2 Detection

The detection is the fact of knowing if the patient in question is victim or not of the hemorrhage, can import the type of it. Thus, it is to classify two classes, the positive that represents the case where the patient is affected, negative in the opposite case. The following table summarize the results given by each of the models in this task (Table 2):

Table 2. Results of the models in the detection task

Model	Loss	Accuracy	Val_Loss	Val_Accuracy
CNN	27.34%	88.69%	26.18%	88.42%
Transfert learning with Inception V3	2.02%	88.44%	28.24%	88.97%

The obtained results in Table 3 on database show that, the CNN model ended up with an accuracy equals to 88.69% and a val_accuracy equals to 88.42%. The values of val_accuracy obtained prove the performance of Transfert learning with Inception V3 and its superiority vis-a-vis CNN.

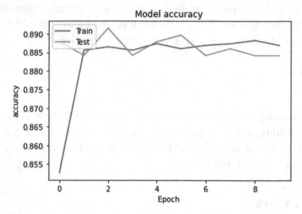

Fig. 4. The accuracy rate of the CNN model in detection task

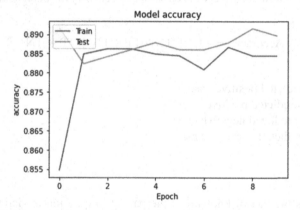

Fig. 5. The accuracy rate of the transfer learning model in detection task

The epochs presented in x-axis is the time consuming when the model is trained on training data or validation data. And accuracy presented in y-axis is the accuracy of the model as a function of the increasing number of epochs.

Two remarks are considered in Figs. 4 and 5. The first one is that the accuracy increases as the number of epochs increases. In the second remark, we observed that during the learning process the difference between the two accuracy curves becomes progressively smaller, which means that is no overfitting problem.

4.3 Classification

After the detection of the hemorrhage, we propose in this part to go further in order to diagnose the type of hemorrhage (epidural, subdural, subarachnoid, Intraparenchymal or intraventricular) by using the two models (CNN model and Transfer Learning). The following table summarize the results given by each of the models in classification task:

Table 3. Results of the models in the classification task

Model	Loss	Accuracy	Val_Loss	Val_Accuracy
CNN	55.72%	51.18%	52.47%	54.69%
Transfer learning with Inception V3	69.32%	78.27%	69.32%	78.75%

Table 4 shows the results of the different models on the training set and the test set in the classification task. The transfer learning model outperformed the other model by an accuracy equal to 78.27% and a val_accuracy equal to 78.75%.

Loss is applied in training data, its values equal to the predicted value after training minus the true value.

Val_Loss is calculated as Loss, but the only difference being that this metric is used on the validation data.

Accuracy is calculated as the ratio between the number of correct predictions to the total number of predictions. This metric is used on the training data.

val_accuracy is calculated as Accuracy, but the only difference being that the metric is used on the validation data.

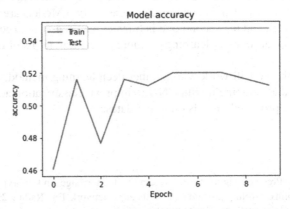

Fig. 6. The accuracy rate of the CNN model in classification task

As shown in Figs. 6 and 7, we did not find the best accuracy numbers due to the fact that we did not have enough data, at the beginning there were only two classes and now we have five classes, which is the reason why the results found are lower than the results of the first task.

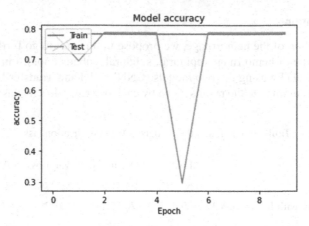

Fig. 7. The accuracy rate of the Transfer Learning model in classification task

The obtained performances are relatively acceptable considering the small size of the data. This is explained by the fact that the convolution neural network is more performing in the case of large data sizes.

In the second part, there is a decrease in accuracy results as a consequence of the change of task and the lack of data, which justifies the existence of overfitting.

If we can obtain a larger database in the future, our method will give better results.

5 Conclusion and Perspectives

In this paper, a comparative study based on Deep learning models to detect and classify the intracranial hemorrhage's database are presented. Two Models are treated, the first one is a CNN model, and the second one is transfer learning. The experiments showed that the second model (transfer learning) is more efficient in terms of training cost and accuracy rate.

In future work, we will work with another deep learning method, RNN (recurrent neural network), and combine it with CNN to improve classification accuracy. We also plan to trained the proposed models on larger datasets.

References

1. Ye, H., et al.: Precise diagnosis of intracranial hemorrhage and subtypes using a three-dimensional joint convolutional and recurrent neural network. Eur. Radiol. **29**(11), 6191–6201 (2019). https://doi.org/10.1007/s00330-019-06163-2
2. Kuo, W., Häne, C., Mukherjee, P., Malik, J., Yuh, E.L.: Expert-level detection of acute intracranial hemorrhage on head computed tomography using deep learning. Proc. Natl. Acad. Sci. **116**(45), 22737–22745 (2019)
3. Tekouabou, S.C.K., Hartini, S., Rustam, Z., Silkan, H., Agoujil, S.: Improvement in automated diagnosis of soft tissues tumors using machine learning. Big Data Min. Anal. **4**(1), 33–46 (2021)

4. Dawud, A.M., Yurtkan, K., Oztoprak, H.: Application of deep learning in neuroradiology: brain haemorrhage classification using transfer learning. Comput. Intell. Neurosci. **2019** (2019)
5. Burduja, M., Ionescu, R.T., Verga, N.: Accurate and efficient intracranial hemorrhage detection and subtype classification in 3D CT scans with convolutional and long short-term memory neural networks. Sensors **20**(19), 5611 (2020)
6. Toğaçar, M., Cömert, Z., Ergen, B., Budak, Ü.: Brain hemorrhage detection based on heat maps, autoencoder and CNN architecture. In: 2019 1st International Informatics and Software Engineering Conference (UBMYK), pp. 1–5. IEEE, November 2009
7. Goodfellow, I., Bengio, Y., Courville, A.: Deep Learning. MIT Press, London (2016)
8. Ker, J., Singh, S.P., Bai, Y., Rao, J., Lim, T., Wang, L.: Image thresholding improves 3-dimensional convolutional neural network diagnosis of different acute brain hemorrhages on computed tomography scans. Sensors **19**(9), 2167 (2019)
9. Castro, J.S., Chabert, S., Saavedra, C., Salas, R.F.: Convolutional neural networks for detection intracranial hemorrhage in CT images. In: CRoNe, pp. 37–43 (2019)
10. Lewick, T., Kumar, M., Hong, R., Wu, W.: Intracranial hemorrhage detection in CT scans using deep learning. In: 2020 IEEE Sixth International Conference on Big Data Computing Service and Applications (BigDataService), pp. 169–172. IEEE, August 2020
11. Nguyen, N.T., Tran, D.Q., Nguyen, N.T., Nguyen, H.Q.: A CNN-LSTM architecture for detection of intracranial hemorrhage on CT scans. arXiv preprint arXiv:2005.10992 (2020)
12. Chilamkurthy, S., et al.: Deep learning algorithms for detection of critical findings in head CT scans: a retrospective study. Lancet **392**(10162), 2388–2396 (2018)
13. Tékouabou, S.C.K., Chabbar, I., Toulni, H., Cherif, W., Silkan, H.: Optimizing the early glaucoma detection from visual fields by combining preprocessing techniques and ensemble classifier with selection strategies. Expert Syst. Appl. **189**, 115975 (2022)
14. Long, M., Cao, Y., Wang, J., Jordan, M.: Learning transferable features with deep adaptation networks. In International conference on machine learning. In: PMLR, pp. 97–105. 2015, June
15. Krizhevsky, A., Sutskever, I., Hinton, G.E.: Imagenet classification with deep convolutional neural networks. Adv. Neural. Inf. Process. Syst. **25**, 1097–1105 (2012)
16. Cheng, P.M., Malhi, H.S.: Transfer learning with convolutional neural networks for classification of abdominal ultrasound images. J. Digit. Imaging **30**(2), 234 (2017)
17. Lei, H., et al.: A deeply supervised residual network for HEp-2 cell classification via cross-modal transfer learning. Pattern Recogn. **79**, 290–302 (2018)
18. Szegedy, C., Vanhoucke, V., Ioffe, S., Shlens, J., Wojna, Z.: Rethinking the inception architecture for computer vision. In: Proceedings of the IEEE Conference on Computer Vision and Pattern Recognition, pp. 2818–2826 (2016)
19. Szegedy, C., et al.: Going deeper with convolutions. In: Proceedings of the IEEE Conference on Computer Vision and Pattern Recognition, pp. 1–9 (2015)
20. Shin, H.C., et al.: Deep convolutional neural networks for computer-aided detection: CNN architectures, dataset characteristics and transfer learning. IEEE Trans. Med. Imaging **35**(5), 1285–1298 (2016)
21. Kumar, A., Kim, J., Lyndon, D., Fulham, M., Feng, D.: An ensemble of fine-tuned convolutional neural networks for medical image classification. IEEE J. Biomed. Health Inform. **21**(1), 31–40 (2016)
22. Nguyen, L.D., Lin, D., Lin, Z., Cao, J.: Deep CNNs for microscopic image classification by exploiting transfer learning and feature concatenation. In: 2018 IEEE International Symposium on Circuits and Systems (ISCAS), pp. 1–5. IEEE, May 2018
23. Hssayeni, M.D., Croock, M.S., Salman, A.D., Al-khafaji, H.F., Yahya, Z.A., Ghoraani, B.: Intracranial hemorrhage segmentation using a deep convolutional model. Data **5**(1), 14 (2020)

Business Intelligence and Database

Increasing Student Engagement in Lessons and Assessing MOOC Participants Through Artificial Intelligence

Younes-aziz Bachiri[(✉)] [iD] and Hicham Mouncif [iD]

Laboratory of Innovation in Mathematics, Applications, and Information Technologies,
Sultan Moulay Slimane University, Beni Mellal, Morocco
{younes-aziz.bachiri,h.mouncif}@usms.ma

Abstract. In today's generation of MOOCs, videos are fundamental to the student learning experience. Due to the prominence of video content in MOOCs, production staff and instructional designers invest significant time and resources in creating these videos. With instructional videos, they want to increase student involvement. Without formative assessment, however, actual engagement is hard to quantify. Nonetheless, a large number of students necessitates a larger pool of questions; to address this issue, we considered mixing machine-learning methods with automatic natural language processing in order to expand the number of questions evaluated and ensure their validity. To accomplish this, we implemented a methodology that generates questions automatically from video transcripts. Following each course comes an evaluation issue, which is typically a multiple-choice question designed to measure a student's comprehension of the video's material. machine-generated questions performed comparably to human-generated questions when it came to judging skill and resemblance. Additionally, the findings indicate that the majority of the questions generated improve e-assessment when the new technology is applied.

Keywords: Massive open online courses · Assessment · Automatic question generation · Computer assisted testing · Artificial Intelligence · Machine learning

1 Introduction

The education business is undergoing a digital change, with video serving as a catalyst, enabling subject matter experts to quickly reach global audiences while also making education more accessible and affordable in general. When it comes to increasing flexibility and personalizing each step of the learning journey, video provides limitless opportunities. Students can learn on the go and pause or restart a session at any moment with video (Moos and Bonde 2016). They can simply return to lectures on demand and spend additional time on concepts or content that they may have missed the first time. For busy professionals who work long hours and have family responsibilities, video allows them to choose what to learn and when to study. Additionally, video enables educators to create individualized learning experiences tailored to the unique needs of each student,

© Springer Nature Switzerland AG 2022
M. Fakir et al. (Eds.): CBI 2022, LNBIP 449, pp. 135–145, 2022.
https://doi.org/10.1007/978-3-031-06458-6_11

as opposed to the one-size-fits-all approach common in physical classroom settings. Educators may develop films and lesson plans for a variety of demographic groups, and by using multi-language subtitles, they can reach students worldwide.

Although video has several benefits, it remains mainly passive until students actively interact and take control of their learning progress. Fortunately, there are a variety of ways to incorporate interactive elements into videos. For example, offering connections to further information or incorporating a multi-view feature enables students to experience critical concepts from a variety of angles (Baker 2016). By gamifying lecture routes and allowing learners to choose what occurs next, you can explain how a series of choices can result in diverse results. Additionally, educators can construct game-based learning programs in which students gain points for making correct choices, trophies are awarded for reaching learning goals, and a final score is used to measure performance.

When an educator has access to an individual's viewing behaviors and preferences, such as the type of content they consume and for how long, they gain insight into their interests and the effectiveness of their lessons or tutorials.

For example, if a high number of viewers skip through to the next video before finishing the current one, there are likely ways to improve the engagement of the lesson. Similarly, if the learner has to keep watching or listening to the video over and over again, it could mean that the lesson is hard to follow or isn't getting through to them.

By determining who is watching what, they can also give personalized recommendations for what to study next based on the content the learner is most likely to enjoy. Then educators can sit back and watch as student conversion and retention rates increase (Klašnja-Milićević et al. 2011).

There is no reversing the trend toward video in education. Even if people start to learn in person again, educational providers will keep using video to reach people all over the world with interactive classes that make them want to come back for more.

To help students learn and get the most out of educational videos, it is important to give them tools that help them process information and keep track of their own understanding.

Lawson et al. (2006) investigated the effects of guiding questions on students' comprehension of a social psychology movie. They had students in some sections of the course watch the film without any additional instructions, while others were given eight guiding questions to consider while watching. Students who responded to the guided questions while seeing the film performed much better on a subsequent test. How can we employ speech-to-text technology and automatic question generation to assess and assist students in developing an active learning style?

The rest of this research is organized as follows. In section two, we summarize the pertinent work in this field. In section three, we describe our methods for automatically generating questions from videos. In section four, we present the criteria and metrics used to evaluate the system, with an emphasis on the results of the key selection classification model. Finally, section five summarizes and concludes the paper.

2 Related Works

Fill-in-the-blank questions, often known as cloze exercises, are a simple way to generate questions. The sentence is first tokenized and tagged for part-of-speech, with the identified entity or noun component masked out. The generated questions are identical to the reading passage questions except for one word or phrase. Although fill-in-the-blank questions are frequently used to assess reading comprehension, they might be misleading if the solution is obvious. A previous study employed supervised machine learning to produce fill-in-the-blank questions. The model anonymizes entity markers to paraphrase the missing word from the passage (Hermann et al. 2015). Semi-automatic approaches can also generate questions. Semi-automatic question generation uses human-generated templates and linked database repositories (Rey et al. 2012). The answer is also retrieved from the associated database. If many answers are required, distractors from the database could be chosen at random as erroneous answers. Sherlock is another example of a template-based question and answer generator that uses linked data to generate questions of variable difficulty (Liu and Lin 2014). However, creating a large list of high-quality questions can be mentally taxing and time-consuming. The questions created must likewise follow the templates. Creating a huge dataset of questions is difficult. Bachiri and Mouncif (2020) discussed defining criteria for selecting an acceptable learning management system and incorporating new automatic natural language processing techniques, such as those employed in artificial intelligence, to increase MOOCs' attractiveness. To do this, they made a plugin that automatically asks questions in different languages and turns the course into a fun game with points and badges.

Numerous reviews discuss various strategies for question generation:

Le et al. (2014) conducted a review of the state of the art for developing educational applications that incorporate question generation techniques. They found that, even though there are many ways to make questions automatically, only a few educational systems based on question generation have been used in real classrooms.

Divate and Salgaonkar (2017) examined a variety of automated question generation systems in order to ascertain why automated question generation continues to be attractive to researchers. The emphasis is primarily on alternative approaches and methodologies' analysis and evaluation.

Dhawaleswar Rao and Saha (2020) conducted a comprehensive study of methods for creating multiple-choice questions automatically. They described a general procedure for generating multiple-choice questions automatically. The procedure is divided into six steps.

Amidei et al. (2018, pp. 2013–2018) conducted a review of the AQG assessment methods. Based on a sample of 37 articles, their research found that the growth of systems has not been matched by an increase in the ways to evaluate them.

Kurdi et al. (2020) summarized the AQG community and its activities, discussed current trends and advancements in the field, highlighted recent innovations, and recommended areas for improvement and future possibilities for AQG.

Das et al. (2021) provided an overview of approaches for automatically generating and evaluating questions based on textual and graphical learning resources. The goal of this study is to gather the most up-to-date methods for making and analyzing questions on your own.

3 Methods: Automatic Question Generation from Videos

Most of the world—including Morocco—has seen a huge increase in demand for instructors and educational tools. Teachers are well aware of the stress of not only grading quizzes but also creating them. At least once a week, the designer of the MOOC must review the videos for the current chapter in order to formulate appropriate questions for the students' weekly tests. Because this procedure can take hours, we investigated ways to automate it for this project, as shown in Fig. 1. In this section, we provide a tool that can help you come up with questions for video material.

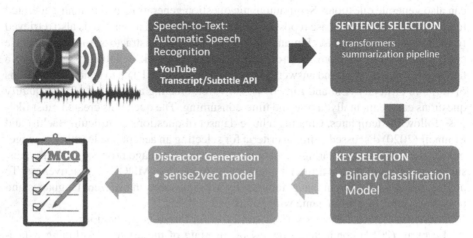

Fig. 1. The proposed system architecture

3.1 Speech-to-Text: Automatic Speech Recognition

With the number of Automatic Speech Recognition (ASR) systems growing at a breakneck pace, it's becoming increasingly difficult to identify the most suited system for a given application. The model takes as input a speech signal and converts it to a series of text-based words. Throughout the voice recognition process, a list of probable texts is generated, and finally, the text that is the most pertinent to the original sound signal is chosen.

The WER, Hper, and Rper error measures are used to compare three ASR systems: IBM Watson, Google, and Wit. Among the three systems tested, Google's automatic voice recognition was found to be the most accurate and efficient (Filippidou and Moussiades 2020).

An API call made by YouTube's browser client allows us to get up the transcript and subtitles for any given video on the site. Subtitle translations are also supported, as well as subtitles that are generated automatically.

The following is an example of the dictionaries returned by the API:

```
[{    'text': 'Hello my dear students',
      'start': 0.58,
      'duration': 1.13
 },{
      'text': ' Now, I'm going to kick off today's lesson with ',
      'start': 2.15,
      'duration': 0.58 },   # ...]
```

YouTube includes a tool that allows you to translate subtitles automatically. Additionally, this module enables access to this capability.

By default, this module always prefers manually made transcripts over computer generated ones when both are available in the desired language.

3.2 Sentence Selection

Summarization is the process of condensing a document while retaining its essential content. Certain models are capable of extracting text from the original input, while others are capable of creating wholly new text (Gupta et al. 2021).

Hugging Face's (HF) transformers summarization pipeline (Wolf et al. 2020), in particular, facilitated the task's execution by making it easier, faster, and more efficient. To be sure, present outcomes are still somewhat hit-or-miss. It is an open-source project that provides numerous valuable natural language processing modules and datasets. Its most well-known library is the Transformer collection. The transformer library includes a variety of pre-trained models for predicting text summaries that may be customized for any dataset.

3.3 Key Selection

It was necessary to train a binary classification model using one of the popular methods such as Logistic Regression, k-Nearest Neighbors, Decision Trees, Support Vector Machine, or Naive Bayes to determine if a word might be used as a response to our generated question (Hoshino and Nakagawa 2005; Goto et al. 2010).

The model is trained using the Stanford Question Answering Dataset. Around 100,000 questions were produced from Wikipedia articles from the SQUAD v1 database (Rajpurkar et al. 2016).

We utilized spacy to describe the characteristics of the words. Once the text has been tokenized, we may access characteristics such as part of speech.

After extracting all the words from the texts, we used pickle to store them. Then we can use them in other modules without waiting for them to regenerate.

As shown in Table 1, we present a sample of the training data set. The total number of words in all articles is up to 8678.

The proportion of extracted words that correspond to responses in the dataset: There are 383 total answers, accounting for 4.41% of all terms.

We encoded the categorical data-NER, POS, TAG, DEP, and shape-and omitted columns containing non-word characteristics.

We utilized the Quadratic Discriminant Analysis classifier to determine whether a word is an answer.

Table 1. A quick overview of the used columns from datasets

TEXT	ISANSWER	NER	POS	TAG	DEP
Architecturally	False	None	ADV	RB	advmod
School	False	None	NOUN	NN	nsubj
Catholic	False	NORP	None	None	None
The Main Building	True	FAC	None	None	None
Saint Bernadette	True	PERSON	None	None	None

3.4 Question Generation

We used the Text2Text model. Based on the open-source PyTorch code for UniLM, a simple-to-use text question and summary generator. Both GitHub and the [Python Packaging Index (PyPI)] host the code. That means it can be readily installed via pip.

The current state of the art in AI and NLP, more specifically natural language generation (NLG), can provide chances to enhance and serve the educational needs of our global community. PyTorch has shown to be an extremely powerful and effective framework for generating and sharing these models and tools, having been adopted by a large portion of the machine learning community (Dong et al. 2019). As illustrated in the Fig. 2, we provide the model with two inputs: the sentence and the target word, and he is then able to construct the question.

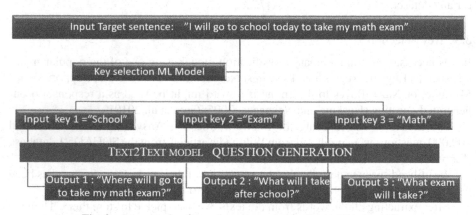

Fig. 2. An example of questions generated by the text2text model

3.5 Distractor Generation

MCQs (Multiple Choice Questions) are frequently used in student evaluations. The question is created, along with a few incorrect responses (referred to as distractors) and the correct answer. QG systems can employ WordNet, internet searches, and Word Sense Disambiguation for assessment.

We employed sense2vec (Trask et al. 2015), a clever variation on word2vec that enables the learning of more interesting and detailed word vectors. This package implements a straightforward Python API for loading, querying, and training sense2vec models.

For each right answer, we took the word and sent it to Sense2Vec to generate the most similar sentences. We disregard the highest-scoring commonalities in order to generate the distractors.

4   Results and Discussion

Figure 3 depicts an illustration of one of the comprehension questions used in this study. They are based on the video question from the USMS MOOC. Each component of a question is as follows: (1) a target word that corresponds to the word being tested in the question, (2) a transcript generated from video footage containing the target word, (3) a correct answer, and (4) distractors. This type of question is designed to assess a test taker's ability to comprehend the meaning of a target word, as well as his or her ability to watch video. Among the four options, there is only one correct answer.

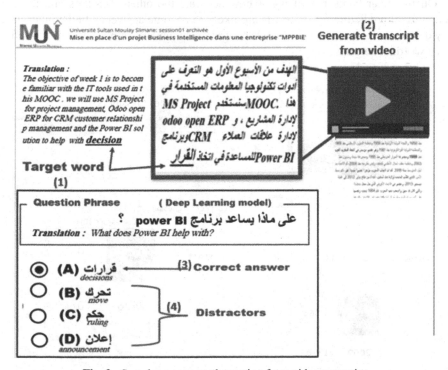

Fig. 3. Sample: a generated question from video transcript

It is self-evident that every word in an informative sentence cannot function as a key. Thus, key selection is a critical step because it determines which word or phrase in the selected sentence will be the key. The following sections discuss the result of the models used.

When evaluating a model for key selection, we consider accuracy but also how robust the model is, how it performs on different datasets, and how much flexibility it offers.

As illustrated in Fig. 4, our classification problem's summary of prediction results. The total number of right and bad predictions is summarized and divided down by class using count values.

The Gaussian naive Bayes classifier has the lowest accuracy of any classifier ever measured, 59.83%, followed by quadratic discriminant analysis (69% accuracy). Other classifiers, such as Linear Discriminant Analysis, Ada Boost Classifier, Random Forest Classifier, Decision Tree Classifier, Support Vector Classification (SVC), K Neighbors Classifier, Decision Tree Classifier, multi-layer perceptron, multinomial, and Bernoulli Naive Bayes, achieve the highest accuracy (91.2%).

While accuracy is a critical statistic to evaluate, it is not always sufficient. Due to the massive mathematical processes required and the uncertain nature of the data, the model may achieve higher accuracy but fails to correctly realize the data, resulting in poor performance when the data is varied. This indicates that the model is not sufficiently robust, hence limiting its application.

On the other hand, regardless of how accurate the other classifiers are, they are incapable of predicting which words are likely to become answers. On the other hand, they discover irrelevant terms, which is natural given that we cannot ask questions about every word in the transcript.

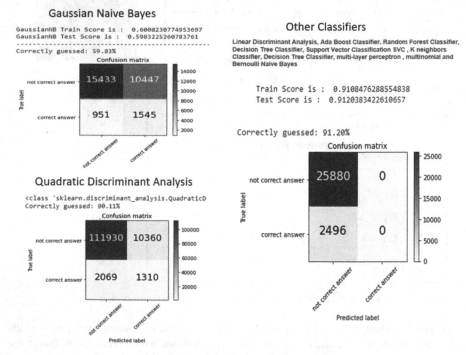

Fig. 4. Key selection confusion matrix

As illustrated in Table 2, accuracy can be deceptive. Occasionally, it may be advantageous to use a model with poorer accuracy if it has a higher predictive potential for the situation.

Recall is a metric that indicates how well our model performs when all the observed values are positive.

If Recall equals 0, the model is defective and is incapable of making even one right prediction when the actual answer is 'yes' (positive). Additionally, this demonstrates that evaluating a model just on the basis of accuracy is not the optimal approach.

The purpose of a research into automatic question generation should not be to produce an endless number of questions. It is vital to guarantee that the questions generated are of a high quality.

So far, we've looked at metrics that, by themselves, cannot adequately represent the model's resilience, but there is one indicator that comes close.

According to (Chicco and Jurman 2020), Matthew's correlation coefficient is the most informative parameter for any binary classifier.

Table 2. Metrics to evaluate our machine learning algorithms

	Gaussian N Bayes	Quadratic discriminant analysis	Other classifiers
Accuracy:	0.598	0.901	0.912
Precision score:	0.129	0.112	0.000
Recall score:	0.619	0.388	0.000
F1 score	0.213	0.174	0.000
Matthew's MCC	0.123	0.169	0.000

Rather than that, the Matthews correlation coefficient (MCC = 16.9%) used in Quadratic Discriminant Analysis is a more reliable statistical rate that delivers a high score only when the forecast performs well in all four confusion matrix sectors.

Text-to-Text models are trained to discover the relationship between two texts (e.g. translation from one language to another). T5, T0, and BART are the most common versions of these devices (Raffel et al. 2020; Lewis et al. 2019). These models are taught to perform a variety of tasks, including summarization, translation, and question generation.

Dong et al. (2019) fine-tuned UNILM on the training set for ten epochs on the SQuAD question generator BLEU-4 to 22.12 (3.75 absolute improvement). They chose 32 as the batch size, 0.7 as the masking probability, and 2e−5 as the learning rate. Label smoothing occurs at a rate of 0.1. The remaining hyper-parameters were identical to those used during pre-training.

5 Conclusion

We have presented an automated technique for creating multiple-choice questions with one correct answer and three distracters. The proposed technique identifies essential phrases with the use of binary classification models. We examined combining machine-learning techniques with autonomous natural language processing to increase the number of questions analyzed while maintaining their validity. To accomplish this, we developed a system for automatically generating questions from video transcripts. Each course concludes with an evaluation question, which is often a multiple-choice question that assesses a student's comprehension of the video's content. To rate skill and similarity, machine-generated questions did just as well as questions that were written by humans.

References

Amidei, J., Piwek, P., Willis, A.: Evaluation methodologies in automatic question generation 2013–2018. In: Proceedings of the 11th International Conference on Natural Language Generation, pp. 307–317 (2018). http://oro.open.ac.uk/57517/

Bachiri, Y., Mouncif, H.: Applicable strategy to choose and deploy a MOOC platform with multilingual AQG feature. In: 2020 21st International Arab Conference on Information Technology (ACIT), pp. 1–6 (2020). https://doi.org/10.1109/ACIT50332.2020.9300051

Baker, A.: Active learning with interactive videos: creating student-guided learning materials. J. Libr. Inf. Serv. Dist. Learn. 10(3–4), 79–87 (2016). https://doi.org/10.1080/1533290X.2016.1206776

Dhawaleswar Rao, C.H., Saha, S.K.: Automatic multiple choice question generation from text: a survey. IEEE Trans. Learn. Technol. 13(1), 14–25 (2020). https://doi.org/10.1109/TLT.2018.2889100

Chicco, D., Jurman, G.: The advantages of the Matthews correlation coefficient (MCC) over F1 score and accuracy in binary classification evaluation. BMC Genomics 21(1), 6 (2020). https://doi.org/10.1186/s12864-019-6413-7

Das, B., Majumder, M., Phadikar, S., Sekh, A.A.: Automatic question generation and answer assessment: a survey. Res. Pract. Technol. Enhanc. Learn. 16(1), 1–15 (2021). https://doi.org/10.1186/s41039-021-00151-1

Divate, M., Salgaonkar, A.: Automatic question generation approaches and evaluation techniques. Curr. Sci. 113(9), 1683–1691 (2017)

Dong, L., et al.: Unified language model pre-training for natural language understanding and generation. Adv. Neural Inf. Process. Syst. 32. https://proceedings.neurips.cc/paper/2019/hash/c20bb2d9a50d5ac1f713f8b34d9aac5a-Abstract.html

Filippidou, F., Moussiades, L.: A benchmarking of IBM, Google and Wit automatic speech recognition systems. In: Maglogiannis, I., Iliadis, L., Pimenidis, E. (eds.) AIAI 2020. IAICT, vol. 583, pp. 73–82. Springer, Cham (2020). https://doi.org/10.1007/978-3-030-49161-1_7

Goto, T., Kojiri, T., Watanabe, T., Iwata, T., Yamada, T.: Automatic generation system of multiple-choice cloze questions and its evaluation. Knowl. Manag. E-Learn. 2(3), 210–224 (2010)

Gupta, A., Chugh, D., Anjum, Katarya, R.: Automated news summarization using transformers (2021). arXiv:2108.01064. http://arxiv.org/abs/2108.01064

Hermann, K.M., et al.: Teaching machines to read and comprehend. Adv. Neural Inf. Process. Syst. 28. https://proceedings.neurips.cc/paper/2015/hash/afdec7005cc9f14302cd0474fd0f3c96-Abstract.html

Hoshino, A., Nakagawa, H.: A real-time multiple-choice question generation for language testing: a preliminary study. In: Proceedings of the Second Workshop on Building Educational Applications Using NLP, pp. 17–20 (2005)

Klašnja-Milićević, A., Vesin, B., Ivanović, M., Budimac, Z.: E-Learning personalization based on hybrid recommendation strategy and learning style identification. Comput. Educ. **56**(3), 885–899 (2011). https://doi.org/10.1016/j.compedu.2010.11.001

Kurdi, G., Leo, J., Parsia, B., Sattler, U., Al-Emari, S.: A systematic review of automatic question generation for educational purposes. Int. J. Artif. Intell. Educ. **30**(1), 121–204 (2019). https://doi.org/10.1007/s40593-019-00186-y

Lawson, T.J., Bodle, J.H., Houlette, M.A., Haubner, R.R.: Guiding questions enhance student learning from educational videos. Teach. Psychol. **33**(1), 31–33 (2006). https://doi.org/10.1207/s15328023top3301_7

Le, N.-T., Kojiri, T., Pinkwart, N.: Automatic question generation for educational applications–the state of art. In: Advanced Computational Methods for Knowledge Engineering, pp. 325–338. Springer, Cham (2014). https://doi.org/10.1007/978-3-319-00293-4

Lewis, M., et al.: BART: denoising sequence-to-sequence pre-training for natural language generation, translation, and comprehension. arXiv:1910.13461 *[Cs, Stat]* http://arxiv.org/abs/1910.13461

Liu, D., Lin, C.: Sherlock: a semi-automatic quiz generation system using linked data. In: International Semantic Web Conference (Posters & Demos), pp. 9–12 (2014)

Moos, D.C., Bonde, C.: Flipping the classroom: embedding self-regulated learning prompts in videos. Technol. Knowl. Learn. **21**(2), 225–242 (2015). https://doi.org/10.1007/s10758-015-9269-1

Raffel, C., et al.: Exploring the limits of transfer learning with a unified text-to-text transformer. arXiv:1910.10683 *[Cs, Stat]*. http://arxiv.org/abs/1910.10683

Rajpurkar, P., Zhang, J., Lopyrev, K., Liang, P.: SQuAD: 100,000+ Questions for Machine Comprehension of Text. arXiv:1606.05250 *[Cs]*. http://arxiv.org/abs/1606.05250

Rey, G.Á., et al. : Semi-automatic generation of quizzes and learning artifacts from linked data. In: Linked Learning 2012: 2nd International Workshop on Learning and Education with the Web of Data, at the World Wide Web Conference 2012 (WWW 2012), Lyon, France (2012). http://lile2012.linkededucation.org/

Trask, A., Michalak, P., Liu, J.: sense2vec—A Fast and Accurate Method for Word Sense Disambiguation In Neural Word Embeddings (2015). arXiv:1511.06388 *[Cs]*. http://arxiv.org/abs/1511.06388

Wolf, T., et al. : Hugging Face's transformers: state-of-the-art natural language processing (2020). arXiv:1910.03771 *[Cs]*. http://arxiv.org/abs/1910.03771

Mining Frequents Itemset and Association Rules in Diabetic Dataset

Youssef Fakir(✉) ⓘ, Abdelfatah Maarouf ⓘ, and Rachid El Ayachi

Faculty of Sciences and Technics, Sultan Moulay Slimane University, Beni Mellal, Morocco
info.dec07@yahoo.fr, rachid.elayachi@usms.ma

Abstract. Data mining is a field of science to extract and analyses the information from large dataset. One of the most techniques is association rule mining. It aim is to find the relationship between the different attributes of data. Several algorithms for extracting data have been developed. Among the existing algorithms the FP-Growth algorithm is one of well-know algorithm in finding out the desired association rules. The aim of this paper is the extraction of association rules by FP-Growth algorithm and its variants using a diabetic dataset, which are the CFP-Growth and ICFP-Growth. Experimental results show that the ICFP-Growth is more accurate than CFP-Growth and FP-Growth.

Keywords: Data mining · Association rules · Frequent patterns · FP-Growth · CFP-Growth · ICFP-Growth

1 Introduction

Extraction of knowledge form large databases is important items in datamining. During the past decades several algorithms have been developed [1–6]. In this paper, we are interested in the association rules algorithms, especially the FP-Growth [7, 8] based algorithms and its variants such as CFP [9, 10] and ICFP-Growth [11–13], by making a comparative study between these three algorithms. The ICFP is an improved version of the CPF-growth algorithm which consist of three steps: the construction of the Multiple Item Support Tree (MIS-Tree) [14, 15], the extraction of the compact MIS-tree and mining the compact MIS-tree. The three algorithms FP-growth, CFP-Growth and ICFP-growth are implemented in order to compare their performances using Python 3 as programming language, vs code as IDE and windows 10 machine with 1.8 GHz and 8 GB memory as environment. The dataset is the female's diabetes dataset (https://www.kaggle.com/mathchi/diabetes-data-set). It is divided into two '.csv' files the first for train dataset, and the other for the test dataset. The two '.csv' files contain 8 features:

- Pregnancies: Number of times pregnant.
- Glucose: Plasma glucose concentration 2 h in an oral glucose tolerance test.
- Blood Pressure: Diastolic blood pressure (mm Hg).
- Skin Thickness: Triceps skin fold thickness (mm).
- Insulin: 2-h serum insulin (mu U/ml).

© Springer Nature Switzerland AG 2022
M. Fakir et al. (Eds.): CBI 2022, LNBIP 449, pp. 146–157, 2022.
https://doi.org/10.1007/978-3-031-06458-6_12

- BMI: Body mass index (weight in kg/(height in m)^2).
- Diabetes Pedigree Function: Diabetes pedigree function.
- Age: Age (years).

The paper is organized as follows. Section 2 describes the processes of database transformation. Section 3 deals with the extraction rules extraction. Performances evaluation of the algorithm are given in Sect. 4 while Sect. 5 concludes the paper.

2 Database Transformation

The dataset contains just numerical values. The FP-Growth, CFP-Growth and ICFP-Growth accept the transactional datasets. The diabetes dataset (numerical datasets) is transformed into a transactional dataset. To do this transformation each feature is visualized in order to know how it change next to the number of individuals and to divide each feature into domains that regroups several individuals. The first feature is the age; the result of visualization is illustrated in Fig. 1.

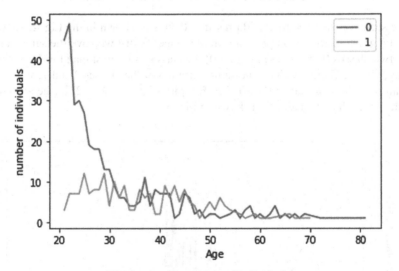

Fig. 1. Age Vs number of individuals

As we can see in the range [20, 30] we have, high number of no diabetes in comparison to the numbers of individuals with diabetes, and for the range [30, 80] we have almost the same number of individuals for both classes 0 and 1 (0: without diabetes, 1: with diabetes), so we can just divide the range of the feature into two domains: A1: [0, 30] and A2: [30, 80].

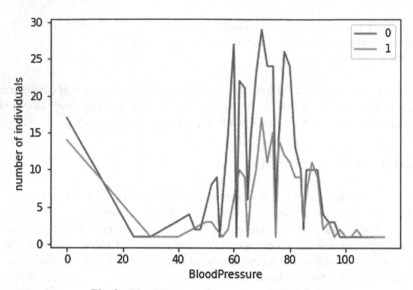

Fig. 2. Blood Pressure Vs number of individuals

The second feature is the blood pressure, the result shown in the Fig. 2. As we can see in the graph of the blood pressure, in the range [0, 40] we have the same variation for the two classes 0 and 1, and in [40, 90] the class 0 is highest than the 1 class, and in the range [90, 120] also we have the same variation of the classes 0 and 1, so we divide this feature to three domains: B1: [0, 40]; B2: [40, 90]; B3: [90, 120]. The third feature is BMI, the result of visualization shown in Fig. 3.

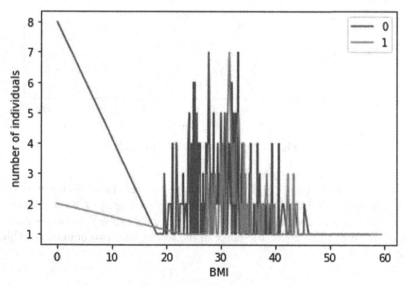

Fig. 3. BMI vs number of individuals

As we see in this graph, we can divide the range of the BMI feature into two domains the first BMI1: [0, 30], where we have individuals' number of the class 0 highest than the 1 class, and the second is BMI2: [30, 60] where the two classes have almost the same variation.

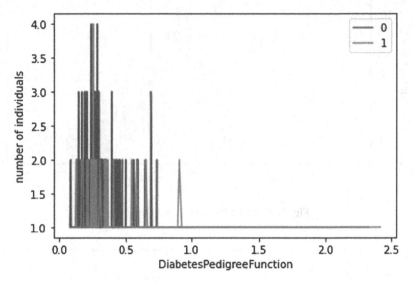

Fig. 4. Diabetes Pedigree Function vs number of individuals

The fourth feature is the Diabetes Pedigree Function, the visualization is in the Fig. 4. In this figure we can see in [0, 0.8] the 0 class have almost the highest number of individuals than the 1 class, and for the range [0.8, 2.5] the opposite, the class 1 have the highest number of individuals, therefore we can divide the feature into two domains: D1: [0, 0.8] and D2: [0.8, 2.5]. The fifth feature is the Glucose, the result of visualization illustrated in the Fig. 5.

As we can see in the range [0, 125] we have, high number of the class 0 in comparison to the numbers of individuals of the class 1, and for the range [125, 200] we have almost the same number of individuals for both classes 0 and 1, so we can divide the range of the feature into two domains: G1: [0, 125] and G2: [125, 200].

The sixth feature is the Insulin, the result shown in the Fig. 6. In the graph of the insulin vs number of individuals in the range [0, 30] we have almost the same variation for the two classes 0 and 1, and in [30, 150] the class 0 is highest than the 1 class, and in the range [150, 800] also we have the same variation of the classes 0 and 1, so we divide this feature to three domains: I1: [0, 30]; I2: [30, 150]; I3: [150, 800]. The seventh feature is the Pregnancies, the result of visualization is shown in the Fig. 7.

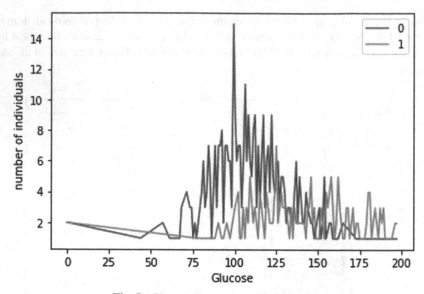

Fig. 5. Glucose Vs number of individuals

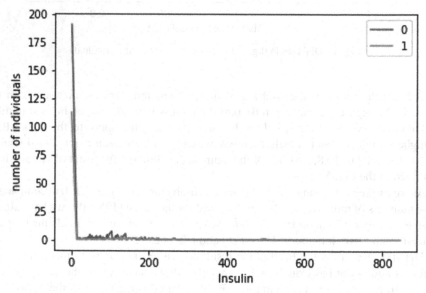

Fig. 6. Insulin Vs number of individuals

As we see in this graph, we can divide the range of the Pregnancies feature into two domains the first P1: [0, 7], where we have individuals' number of the class 0 highest than the 1 class, and the second is P2: [7, 17] where the two classes have almost the same variation. The last feature is the Skin Thickness, the visualization is in the Fig. 8.

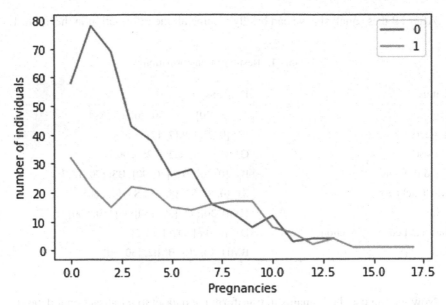

Fig. 7. Pregnancies Vs number of individuals

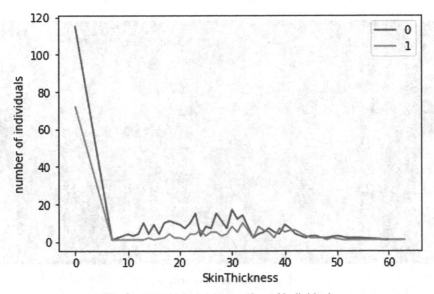

Fig. 8. Skin thickness Vs number of individuals

In this figure we can see in [0, 8], the proximately the same variation for the two classes, and in [8, 45] the 0 class have almost the highest number of individuals than the 1 class. In addition, for the range [45, 60] also we have the same variation for the both classes, therefore we can divide the feature into three domains: S1: [0, 8], S2: [8, 45] and S3: [45, 60].

After all these analysis, we can briefly resume all the information in the Table 1.

Table 1. Result of transformation

Feature	Domains
Age	A1: [0, 30]; A2: [30, 80]
Pregnancies	P1: [0, 7]; P2: [7, 17]
Glucose	G1: [0, 125]; G2: [125, 200]
Blood Pressure	B1: [0, 40]; B2: [40, 90]; B3: [90, 120]
Skin Thickness	S1: [0, 8]; S2: [8, 45]; S3: [45, 60]
Insulin	I1: [0, 30]; I2: [30, 150]; I3[150, 800]
Diabetes Pedigree Function	D1: [0, 0.8]; D2: [0, 2.5]
BMI	BMI1: [0, 30]; BMI2: [30, 60]

Now we can use the domains to transform the dataset to a transactional dataset. As we can see in the Fig. 9, we have a part of result that we obtained from the transformation.

```
[['P1', 'G1', 'B2', 'S2', 'I2', 'BMI2', 'D1', 'A1', '0'],
 ['P2', 'G2', 'B3', 'S2', 'I2', 'BMI2', 'D1', 'A2', '1'],
 ['P1', 'G1', 'B2', 'S2', 'I3', 'BMI2', 'D1', 'A1', '1'],
 ['P1', 'G1', 'B2', 'S3', 'I3', 'BMI2', 'D2', 'A2', '0'],
 ['P1', 'G2', 'B2', 'S2', 'I3', 'BMI2', 'D1', 'A1', '1'],
 ['P1', 'G1', 'B2', 'S1', 'I1', 'BMI1', 'D1', 'A2', '0'],
 ['P1', 'G1', 'B1', 'S2', 'I1', 'BMI1', 'D1', 'A1', '0'],
 ['P1', 'G1', 'B2', 'S3', 'I2', 'BMI2', 'D1', 'A1', '0'],
 ['P1', 'G1', 'B2', 'S2', 'I2', 'BMI2', 'D1', 'A1', '0'],
 ['P1', 'G1', 'B2', 'S2', 'I2', 'BMI2', 'D1', 'A1', '0'],
 ['P1', 'G1', 'B2', 'S2', 'I2', 'BMI1', 'D2', 'A1', '0'],
```

Fig. 9. Part of transformation

3 Extraction of Association Rules

First, we have to initialize the minsupport for FP-Growth [17], and MIS-values for CFP-Growth and ICFP-Growth. To assign MIS-values for CFP-Growth, we use the Eq. (1).

$$MIS(i) = maximum(\beta f(i), LS) \tag{1}$$

- MIS (i) is the MIS-value of the item "i".
- $\beta \in [0, 1]$ is a parameter that controls how the MIS values for items should be related to their frequencies.
- $f(i)$ is the frequencies value for the item "i".
- LS is a use specified value, represent the least minimum support allowed.

In addition, for the ICFP-Growth we use the Eq. (2).

$$MIS(i) = \begin{cases} M(i) \ if \ M(i) > LMS \\ LMS \ else \ if \ M(i) < LMS \ and \ S(i) > LMS \\ LMIS \ else \end{cases} \tag{2}$$

with

$$M(i) = S(i) - SD$$

where

- $SD \in [0, 1]$ is a user specified value.
- $S(i) = \frac{f(i)}{N}$ is the support of the item "i".
- $f(i)$ is the frequencies value for the item "i".
- N represent the number of transactions in the dataset.
- LMS is a user specified value, stand for lowest minimum support, and represent lowest MIS value of a frequent item.
- LMIS is also a user specified value, stand for least minimum item support, and represent the lowest MIS value among all items in the transaction dataset.
- The LMIS value should always be less than or equal to LMS.

In this experiment, we define the minsupport of the FP-Growth equal to 40, for the CFP-Growth the β equal to 0.1 and LS equal to 40, and for the ICFP-Growth, the SD value is 0.1, LMS equal to 50 and LMIS equal to 40.

The result of the MIS-values generation, for the CFP-Growth is shown in Table 2. In addition, for ICFP-Growth the result of MIS-values initialization is given in the Table 3.

Table 2. MIS value for CFP-Growth algorithm

item	P1	G1	B2	S2	I2	BMI2	D1	A1	0
MIS-value	51	40	55	40	40	40	53	40	40

P2	G2	B3	A2	1	I3	S3	D2	S1	BMI1	B1
40	40	40	40	40	40	40	40	40	40	40

Table 3. MIS value for ICFP-Growth algorithm

item	P1	G1	B2	S2	I2	BMI2	D1	A1	0
MIS-value	40	40	40	40	40	40	40	40	40

P2	G2	B3	A2	1	I3	S3	D2	S1	BMI1	B1
40	40	40	40	40	40	40	40	40	40	40

After we apply the three algorithms on our dataset, we obtain three models that contains the association rules as shown in Fig. 10.

```
('P1',)==>(('0', 'B2', 'D1'), 0.5750487329434698)
('D1',)==>(('0', 'B2', 'P1'), 0.5555555555555556)
('B2',)==>(('0', 'D1', 'P1'), 0.5373406193078324)
('G1',)==>(('I1',), 0.5026315789473684)
('G1', 'P1')==>(('0', 'B2', 'D1', 'S2'), 0.5259938837920489)
('P2',)==>(('B2', 'I1'), 0.5346534653465347)
('G1', 'P2')==>(('A2', 'B2'), 0.8301886792452831)
('S2',)==>(('0', 'B2', 'D1', 'P1'), 0.5544303797468354)
('G2',)==>(('1',), 0.5769230769230769)
```

Fig. 10. Association rules obtained

The structure of our model is: (left) ➜ (right, Confidence) . The left is the causes and the right is the consequence. Confidence is a number in range of [0, 1] that can represent how much left can lead us to the right, who much the causes can lead as to a consequence, and we can use the Eq. (3) to calculate the confidence [16].

$$Confidence(right \rightarrow left) = \frac{support(right \cup left)}{support(right)} \qquad (3)$$

Figure 11 shows the association rules between all the features, but in our case, we want to do a classification model, for that we filter the association rules to have in the consequences (right) just the items that represent the classes ('0' and '1'), the result is shown in Fig. 11.

```
('A1', 'B3', 'BMI2', 'D1', 'G2', 'I2', 'P1')==>(('0',), 1.0)
('A1', 'BMI2', 'D1', 'G2', 'I2', 'P1', 'S3')==>(('0',), 0.5)
('A1', 'B3', 'BMI2', 'D1', 'G2', 'I2', 'P1', 'S3')==>(('0',), 1.0)
('A2', 'BMI2', 'G2', 'S3')==>(('1',), 1.0)
('A2', 'B3', 'BMI2', 'D1', 'P2')==>(('1',), 0.8571428571428571)
('A2', 'BMI2', 'G2', 'P2')==>(('1',), 0.8857142857142857)
('A2', 'BMI2', 'D1', 'G2', 'S3')==>(('1',), 1.0)
('A2', 'BMI2', 'D1', 'G2', 'P2')==>(('1',), 0.8928571428571429)
('A2', 'I3', 'P2')==>(('1',), 0.9375)
('A2', 'BMI2', 'D1', 'I3', 'P2')==>(('1',), 0.9090909090909091)
('A2', 'BMI2', 'G2', 'I3', 'P2')==>(('1',), 1.0)
('A2', 'BMI2', 'G2', 'I3', 'P2', 'S3')==>(('1',), 1.0)
('A2', 'BMI2', 'D1', 'G2', 'I3', 'P2', 'S3')==>(('1',), 1.0)
('A2', 'B3', 'BMI2', 'D1', 'G2', 'I3', 'P2', 'S3')==>(('1',), 1.0)
```

Fig. 11. Results of association rules

In the figure we have the association for the classification model for example we have this association rule ('A2', 'BMI2', 'G2', 'P2') ➔ (('1'), 0.89).

That mean if the individual has A2 the age between [30, 80]. The BMI2 the Body mass index in range of [30, 60], the G2 Plasma glucose concentration 2 h in an oral glucose tolerance test in range of [125, 200], and the P2 number of times pregnant between [7, 17], so we can see that the individual has a diabetes with the confidence of 0.89.

4   Performances Evaluation

The three algorithms FP-Growth, CFP-Growth and ICFP-Growth are evaluated using the same preprocessing that we were applied on the train dataset, to transform the numerical dataset into a transactional dataset. After that, we take a transaction from the dataset and calculate the distance between the test transaction and the left of the association's rules of the model. In this case, we use an approach to calculate the distance. For example, we have T test transaction and G the left of an association rule that exist in the model.

T = ['P1', 'G1', 'B2', 'S2', 'I3', 'BMI2', 'D2', 'A2'],
G = ['A1', 'B3', 'BMI2', 'D1', 'G2', 'I2', 'P1'].

First, we have eight features in datasets. We initialized the distance by 8, and we check for each item of the T if it exist in the G and for each item exist we decrement the distance by one. In this example, we have P1 exist in the T and G, so the distance is 7. In addition, G1 not exist in G and so we still have distance is 7, and B2, S2, I3, D2, A2 also doesn't exist in G, and we have BMI2 exist so we decrement the distance by 1 so we have the distance equal to 6, the distance between G and T is 6. In addition, after calculating the distances, we chose the three closet association rules, we count who much votes for the '0' and for '1', and we choose the class that have the highest number of votes. After this testing process, we calculate the accuracy for each algorithm. The accuracy for the three algorithms FP-Growth, CFP-Growth and ICFP-Growth respectively is 51.30%, 57% and 60.5%.

5 Conclusion

Frequent itemset mining is an important subject in data mining. In this paper, three association rules algorithms, which are FP-Growth, CFP-Growth, and ICFP-Growth, are implemented. These algorithms are used to extract item sets frequents on a diabetes dataset using python programming language. Experimental results show that the ICFP is more accurate than the others algorithms.

References

1. Han, J., Pei, J., Yin, Y.: Mining frequent patterns without candidate generation. In: Proceedings of the 2000 ACM SIGMOD International Conference on Management of Data - SIGMOD 2000, pp. 1–12. Dallas, Texas, United States (2000)
2. Fakir, Y., et al.: A comparative study between Relim and SaM algorithms. Int. J. Comput. Sci. Inf. Secur. 18(5) (2020)
3. Ya-Han, H., Chen, Y.-L.: Mining association rules with multiple minimum supports: a new mining algorithm and a support tuning mechanism. Decis. Supp. Syst. 42(1), 1–24 (2006)
4. Joshua, J.V., et al.: Data mining: a book recommender system using frequent pattern algorithm. J. Softw. Eng. Simul. 3(3), 01–13 (2016)
5. Pathania, S., Singh, H.: A new associative classifier based on CFP-Growth++ algorithm. In: Proceedings of the Sixth International Conference on Computer and Communication Technology 2015, pp. 20–25. New York, NY, USA (2015)
6. Uday Kiran, R., Krishna Reddy, P.: Novel techniques to reduce search space in multiple minimum supports-based frequent pattern mining algorithms. In: Proceedings of the 14th International Conference on Extending Database Technology - EDBT/ICDT 2011, p. 11. Uppsala, Sweden (2011)
7. Fakir, Y., et al.: Closed frequent itemsets mining based on It-Tree. Global J. Comput. Sci. Theory Res. 11(1) (2021)
8. Fakir, Y., El Ayachi, R., Fakir, M.: Mining frequent pattern by titanic and FP-Tree algorithms. Int. J. Sci. Res. Comput. Sci. Eng. Inf. Technol. 6(5) (2020)
9. Uday Kiran, R., Krishna Reddy, P.: An improved frequent pattern-growth approach to discover rare association rules. In: Proceedings of the International Conference on Knowledge Discovery and Information Retrieval, pp. 43–52. Funchal - Madeira, Portugal (2009)
10. Liu, G., et al.: CFP-tree: a compact disk-based structure for storing and querying frequent itemsets. Inf. Syst. 32(2), 295–319 (2007)
11. Darraba, S., Ergenc, B.: Vertical pattern mining algorithm for multiple support thresholds. In: International Conference on Knowledge Based and Intelligent Information and Engineering Systems, KES2017, 6–8 September 2017, Marseille, France (2017)

12. Ahmed, S.A., Nath, B.: ISSP-tree: an improved fast algorithm for constructing a complete prefix tree using single database scan. Expert Syst. Appl. **185**, 115603 (2021)
13. Gan, W.: Mining of frequent patterns with multiple minimum supports. Eng. Appl. Artif. Intell. **60**, 83–96 (2017)
14. Chen, X., Huang, H., Du, W.: Most frequent item sets mining algorithm based on MIS-Tree and multiple support array. Int. J. Res. Eng. Sci. **4**(7) (2016)
15. Sinthuja, M., Rachel, S.S., Janani, G.: Mis-Tree algorithm for mining association rules with multiple minimum support. Bonfring Int. J. Datamin. **1**, 1–5 (2021)
16. Fakir, Y., El Ayachi, R.: Frequent patterns mining. Int. J. Sci. Res. Comput. Sc. Eng. Inf. Technol. **6**(4) (2020)
17. Narvekara, M., Syed, S.F.: An optimized algorithm for association rule mining using FP tree. In: International Conference on Advanced Computing Technologies and Applications (ICACTA-2015), Procedia Computer Science, vol. 45, 101–110 (2015)

Automatic Text Summarization for Moroccan Arabic Dialect Using an Artificial Intelligence Approach

Kamel Gaanoun[1]([✉])(iD), Abdou Mohamed Naira[1,2](iD), Anass Allak[1,2](iD), and Imade Benelallam[1,2]

[1] Institut National de Statistique et d'Economie Appliquée, SI2M Lab., Rabat, Morocco
{kgaanoun,nabdoumohamed,aallak,i.benelallam}@insea.ac.ma
[2] AIOX LABS, Rabat, Morocco
https://www.aiox-labs.com/

Abstract. A major advantage of artificial intelligence is its ability to automatically perform tasks at a human-like level quickly; this is needed in many fields, and more particularly in Automatic Text Summarization (ATS). Several advances related to this technique were made in recent years for both extractive and abstractive approaches, notably with the advent of sequence-to-sequence (seq2seq) and Transformers-based models. In spite of this, the Arabic language is largely less represented in this field, due to its complexity and a lack of datasets for ATS. Although some ATS works exist for Modern Standard Arabic (MSA), there is a lack of ATS works for the Arabic dialects that are more prevalent on social networking platforms and the Internet in general. Intending to take an initial step toward meeting this need, we present the first work of ATS concerning the Moroccan dialect known as Darija. This paper introduces the first dataset intended for the summarization of articles written in Darija. In addition, we present state-of-the-art results based on the ROUGE metric for extractive methods based on BERT embeddings and K-MEANS clustering, as well as abstractive methods based on Transformers models.

Keywords: Text summarization · Extractive · Abstractive · BERT · T5 · Transformers · Arabic · Moroccan dialect

1 Introduction

With the explosion of data in all its forms and more specifically, the continuous creation of textual content, we have entered a contradictory era. Due to time constraints and the need for a huge amount of human effort, we are not taking full advantage of this boon for the acquisition of information. Automatic Text Summarization (ATS) is a technique that meets this need by proposing different approaches all of which aim to provide a more concise, easy-to-understand version of a given text that includes relevant information. There have been many

M. Fakir et al. (Eds.): CBI 2022, LNBIP 449, pp. 158–177, 2022.
https://doi.org/10.1007/978-3-031-06458-6_13

efforts in this direction, especially for the English language. However, the complexity of the Arabic language and its structural and morphological features has left works relating to ATS to remain scarce [1]. If this observation is true for the Arabic language in its official version, in this case, the Modern Standard Arabic (MSA), it is the same for different dialects. In fact, the Arabic language is spoken by 22 countries in the world and represents over 400 million speakers; however, the MSA version is only used for media and official channels, while local dialects are used for everyday use and for social networking sites [2]. As far as we know, no work has been done on the summarization of Arabic dialect texts. By using the Moroccan Arabic dialect, commonly known as Darija, we propose to take a first step towards resolving this problem. Our aim is to provide researchers with a reference to develop powerful systems of summarization, and that with various contributions that are summarized as follows:

- We introduce and release Moroccan Articles **Sum**marization (**MArSum**) dataset: a first dataset for Moroccan dialect text summarization
- We present and compare first results from both extractive and abstractive summarization tasks on Moroccan dialect
- We release a fine-tuned T5 model for Moroccan dialect summarization

The rest of the paper has the following structure: Sect. 2 provides a brief overview of ATS related work and background, Sect. 3 describes the experimental setup, Sect. 4 presents the results, Sect. 5 highlights important discussions about the results while Sect. 6 concludes the paper.

2 Related Work and Background

There are three main ATS approaches: the extractive, the abstractive, and the hybrid approach combining both. The extractive approach remains the most common and achieves the best results, being the simplest to implement and the oldest approach. The abstractive and hybrid approaches, which are much closer to human summarization, are gaining importance with promising results. It is worth noting that work on ATS is not novel. The first works appeared in the 1950s, specifically the work of Luhn [3], where he analyzed technical papers and magazine articles by extracting statistical information to give a score to each sentence. He then retained the sentences with the highest scores for the final abstract. Since then, works focusing on extractive summarization have continued to explore different statistical analysis techniques based on the concept of "most important sentences" [4], like in [5], where they extract 11 statistical features to score sentences and select most important ones, i.e. with the highest score. Other works, such as [6] and [7], use topic based techniques to perform extractive summarization. These latter techniques and ones similar can be qualified as unsupervised, whereas new supervised techniques are increasingly being used, treating the problem as a case of supervised classification to predict for each sentence whether it represents a summary sentence or not. In addition, and with the explosive availability of data and the exponential improvement of computational resources, neural networks and deep learning in general are revolutionizing this field with increasingly powerful results [8].

Contrary to these works consisting of copying a portion of the text (the most important part) to create a concise version, the abstractive summarization is intended to generate a text taking back the main information contained in the text by using new words and new sentences. Although this method is similar to what humans produce in terms of summarization, it is complex to implement and the results suffer from various problems, including repeated words, out of vocabulary tokens, out of scope summaries, and others. Abstractive summarization approaches can be classified as structure based, semantic based or deep learning based [9]. Deep learning based approaches perform better than the former two; Gupta [9] benchmarked the different techniques on the Document Understanding Conference (DUC) dataset to show that the deep learning based works achieved ROUGE scores of up to 0.47, outperforming the other approaches. The most recent techniques in this direction focus on sequence-to-sequence models using jointly trained models, called Encoder for extracting a representation of the input text, and Decoder for generating a summary text; the differences are essentially based on the nature of the model used on both sides (Encoder and Decoder). If the Recurrent Neural Networks (RNN) [10] and Long Short Term Memory (LSTM) [11,12] were initially used, there is a greater preponderance of models based on Transformers [13], and recently, the use of Bidirectional Encoder Representations from Transformers (BERT) models as Encoder [14,15].

To circumvent the caveats of the abstractive approaches, and to also take advantage of the benefits of extractive approaches, new hybrid methods combining both have been explored, mainly with a copy mechanism initiated by [10] and extended by [16] in their pointer-generator models, calculating a probability to make the decision either to copy part of the original text or to generate a new one from an external vocabulary.

Although there are many works on ATS, they all have one thing in common: they are mostly centered around English or other languages with a large amount of data. Arabic suffers from a scarcity of summarization works, but the existing works have followed the evolution of the methods used in literature. If the first works for the Arabic language were carried out using clustering methods, such as in [17], or by word frequency and other statistical features, as featured in [18,19], then the deep learning approaches receive more attention. This is even more remarkable for works related to abstractive summarization: Wazery [20] used seq2seq architectures with different types of encoder and decoder layers, Suleiman [21] also used a seq2seq based system with LSTM as encoder and decoder layers, whereas Al-Maleh [22] used the pointer-generator model. No published work has dealt with summarization for Arabic dialects, to the authors' knowledge.

Transformer and BERT Models. Before the arrival of Transformer-based models, the work producing the state-of-the-art results mainly used RNN models to process texts sequentially. Next, the LSTM models were used, like in

ELMO [23], to alleviate the "vanishing gradient" phenomenon in order to be able to process long sequences. However, a major problem with these models was that it was not possible to focus on the most important part of the sequences as humans do, which is the principal problem Transformers addressed using a new mechanism described in the original paper [13] called "Attention". The Transformer model is composed of two components: An encoder that produces an embedding representation of the input text, and a decoder that transforms this embedding to a text output. The first application of the Transformer model was machine translation. During the training of the Transformer, instead of using recurrent connections like RNNs, it passes processed text through multiple attention layers (see Fig. 1), called "attention heads", and each layer learns different linguistic information, similar to how Convolutional Neural Networks (CNN) models [24] process images.

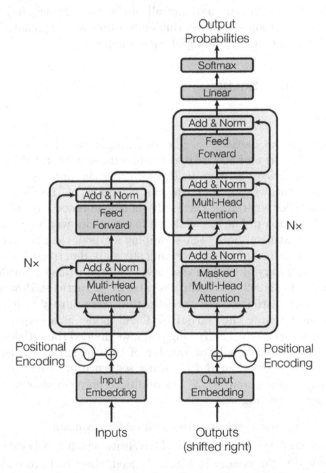

Fig. 1. Transformer model (as described in the original paper [13])

BERT is, as the name suggests, a model based on the encoder part of the Transformer architecture to generate a text representation; it is a language model in terms that it represents contextual relations between language words. Initial BERT models were released by authors [25] in two versions: a BERT Base and a BERT Large. The two models differ in terms of the number of encoder layers, 12 and 24, respectively, which impacts on the remaining components; BERT Base has 12 attention heads and 768 hidden layers whereas BERT Large has 16 attention heads and 1024 hidden layers. One of the advantages of BERT is that it can be pre-trained in a non-supervised scenario on a large corpus and then use transfer learning and fine-tune it on a smaller dataset for specific tasks. Two pre-training tasks are involved: Masked Language Model (MLM) and Next Sentence Prediction (NSP). The first instructs the encoder to learn model language by predicting random masked words, while the latter aims to predict if two sentences follow each other. In addition to embeddings produced by each layer, BERT also adds an embedding layer summarizing all of the embeddings and is generally intended for classification purposes; this embedding is encapsulated in a token called [CLS] added at the beginning of each sentence.

3 Experimental Setup

3.1 Dataset

To our knowledge, no dataset specific to summarization is available for the Moroccan dialect, thus we have created the first dataset intended to work on the summarization of articles written in Darija, using the process shown in Fig. 2. To achieve this goal, we proceeded to the scrapping of the Goud.ma website offering news articles written in Darija. It should be noted that all the titles of the articles are written in Darija, while we have articles written either in Darija or in MSA, or a mixture of both. The scrapping method, the size and the content of the dataset, as well as the preprocessing steps are described bellow.

The Selenium library of Python was used to scrape articles published between 01/01/2018 and 31/12/2020 to obtain a total of 67,115 articles. However, in order to obtain a purely Darija dataset, articles that were mostly written in Darija were filtered out by using the DarijaBERT [26] model as a classifier. However, as BERT is not designed to classify long texts, we divided the articles into several sequences of size 40 tokens with an overlap of 10 tokens, i.e., the last 10 tokens of a sequence are repeated in the following sequence in order to preserve the context, a technique inspired by [27]. In addition, preprocessing steps have also been performed, consisting of:

- Delete the email addresses that the articles may contain
- Replace the word كود (the name of the website which is a Darija word) with the word الجريدة (Newspaper in MSA). Indeed, this word misleads the model during the classification process.

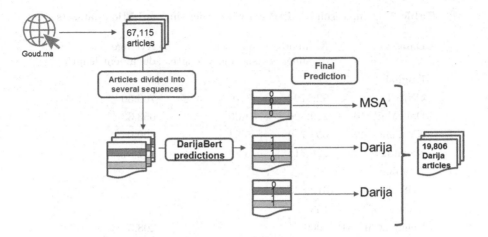

Fig. 2. MArSum dataset creation process

Then DarijaBERT, already fine-tuned on a larger dataset, is used to predict the language of each sequence: coded 1 for Darija and 0 otherwise. In order to determine the predominant language of each article, the average of the predictions was computed over all the sequences of the article, with the idea being that the closer this average is to 1, the more sequences the article contains written in Darija. We estimate after observation that the articles with an average higher than 0.85 were mostly written in Darija and this part of the initial dataset[1] was retained. Table 1 summarizes the characteristics of the dataset that we call MArSum and that is publicly accesible[2].

Table 1. MArSum dataset description.

Size	Titles length			Articles length		
	Minimum	Maximum	Average	Minimum	Maximum	Average
19,806	2	74	14.6	30	2964	140.7

A comparison of MArSum with some other summarization datasets for Arabic and English is included in Table 2 and a sample of the dataset is available in Fig. 8. We note that the dataset was split into a training dataset, reserved for the training process, and a test dataset for evaluation. We used a 90/10 split strategy.

---

[1] We retained articles with at least 30 words.
[2] https://github.com/KamelGaanoun/MoroccanSummarization.

Table 2. Comparison of MArSum with other summarization datasets

Datasets	Number of documents	Average summary length	Average document length
English			
CNN [10]	92,579	45.70	760.50
DailyMail [10]	219,506	54.65	653.33
NY Times [28]	654,759	45.54	800.04
XSum [29]	226,711	23.26	431.07
Arabic			
AHS [22]	312,838	–	–
EASC [30]	153	–	–
Helmy et al. [31]	6000	–	208.5
AMN [32]	265,000	–	–
MArSum	**19,806**	**14.6**	**140.7**

3.2 Extractive Summarization

For the extractive summarization, we adopted the method used by [33] consisting of using the embeddings generated by a BERT model and then, using the K-Means classification, and grouping the n sentences closest to the centroids, as illustrated in Fig. 3. As described by the author, it is a dynamic summarization providing the possibility to specify the number of sentences wanted for the summarized result. The definition of the number of K-classes is done automatically using the Elbow method, then the n closest sentences to the centroids are selected; the number n of sentences is parameterizable hence the dynamic nature of this method. In summary, the summarization is done in three steps: (i) the BERT model generates sentences' embeddings, (ii) those embeddings are clustered using K-Means, and finally (iii) the closest embedded sentences to the centroid are selected.

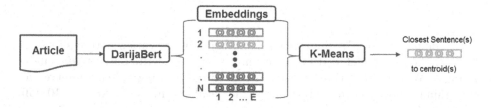

Fig. 3. Extractive summarization process

The efficiency of this method and of the newly created DarijaBERT model (the first Darija-specific BERT model) was used to perform summarization of MArSum articles. The author specifies that embeddings of the second to last hidden layers of the BERT model gave the best results. We adopt the same configuration and experiment with summarization based on a single sentence and then on two sentences. As BERT produces embeddings in a N × W × E dimension (N: number of sentences, W: tokenized words, E: embedding dimension), 4 different aggregation scenarios were run to obtain an N × E embedding matrix and run clustering on it. The aggregations are: mean, median, max and min.

3.3   Abstractive Summarization

For abstractive summarization, we opted for methods based on the sequence-to-sequence [34] architecture. To be more specific, we implemented the method using a transformer architecture. For this, we compared two different approaches: training a model from scratch based on our training dataset, and using transfer learning. For the latter, a T5 model [35] fine-tuned for summarization on the Arabic T5 Model [36] was evaluated (we call it Pre-trained T5 model), T5 [37] being a text-to-text model also based on Transformers architecture. We use the HuggingFace Transformers library [38] to run the inference on our test dataset. Then, the model is further fine-tuned on training data (see Fig. 4) to obtain a model more accurate for Darija text (we call it Fine-tuned T5 model). It is noteworthy to mention that the Pre-trained T5 model was initially fine-tuned on the Arabic and English part of the mC4 dataset, which is based on the Common Crawl's corpus[3].

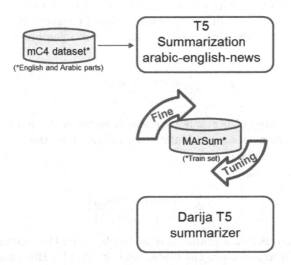

Fig. 4. Abstractive summarization process

Due to a lack of computing resources and for faster training, different hyper-parameters were used than those in the original paper for our from-scratch

---

[3] https://commoncrawl.org.

Transformer model. Hyperparameters were set as follows: num_layers = 4, d_model = 128, dff = 512, num_heads = 4, encoder and decoder maximum length are set respectively to 900 and 120, and we train on the training dataset for 30 epochs. For the Pre-trained T5 model we adopt the following hyperparameters: maximum length = 512, repetition penalty = 1.5, length penalty = 1 and num_beams = 2. Regarding the Fine-tuned T5 model's hyperparameters, they consisted of a learning rate of $1e^{-4}$, maximum source text length of 128, maximum generated summary length of 50, and the training was done for 3 epochs.

3.4   Evaluation

Many evaluation metrics for ATS [39], and for texts similarity in general, were used in the literature, each with its own advantages and disadvantages. However, Recall-Oriented Understudy for Gisting Evaluation (ROUGE) [40] remains the most widely used metric to compare state-of-the-art results. It is a collection of indicators that compares generated summary with the human written ground-truth references. ROUGE indicators include ROUGE-N and ROUGE-L metrics.

ROUGE-N (for unique reference summary) evaluates the n-gram overlap between generated and reference summaries, where for N = 1, ROUGE-1 gives the overlap of unigrams, and for N = 2, ROUGE-2 calculates the overlap between bigrams. ROUGE-N is related to recall measure if it is related to n-grams Located at the reference summary, or as a precision related measure when we compare to n-grams occurring in the generated summary, or as a combination of both using the F-measure.

ROUGE-N$_{Recall}$

$$\frac{\sum_s match(gram_{s,g})}{\sum_s count(gram_s)} \tag{1}$$

In (1) the number of n-grams that match between the reference (s) and the generated (g) summaries are calculated and divided by the number of n-grams in the reference summary.

ROUGE-N$_{Precision}$:

$$\frac{\sum_s match(gram_{s,g})}{\sum_s count(gram_g)} \tag{2}$$

In (2) the number of n-grams that match between the reference (s) and the generated (g) summaries are calculated and divided by the total number of n-grams in the generated summary.

ROUGE-N$_{Fmeasure}$:

$$\frac{\text{ROUGE-N}_{Precision} \times \text{ROUGE-N}_{Recall}}{\text{ROUGE-N}_{Precision} + \text{ROUGE-N}_{Recall}} \tag{3}$$

ROUGE-L evaluates the longest common subsequence (LCS) between the two summaries without requiring consecutive matches but only in-sequence matches.

For evaluation, ROUGE-1, ROUGE-2, and ROUGE-L are reported related to F measure as it summarises the results obtained in terms of precision and recall. The total ROUGE scores are calculated as an average of all articles scores.

Concerning the extractive summarization, as the method used does not require training beforehand, the results reported are obtained on the whole MAr-Sum dataset. As for the abstractive summarization, the results are reported on the test dataset for the two models, and also for the whole dataset for the Pre-trained T5 model.

This said, in order to compare with the abstractive summarization, the results of the extractive summarization restricted to the test dataset are also reported.

4   Results

4.1   Extractive Summarization

The results of the different configurations and aggregation methods over the entire dataset and the test dataset are shown in Table 3 and Table 4; note that all aggregation methods resulted with the same scores for the one sentence summarization.

Table 3. Extractive summarization ROUGE F1 scores on MArSum dataset with different sentence embedding aggregations.

	One sentence	Two sentences			
		Mean	Median	Min	Max
ROUGE-1	26.77	21.62	21.37	21.62	22.38
ROUGE-2	8.25	5.45	5.33	5.45	5.76
ROUGE-L	24.85	20.12	19.81	20.12	20.67

Table 4. Extractive summarization ROUGE F1 scores on test dataset.

	One sentence	Two sentences
ROUGE-1	29.14	20.68
ROUGE-2	5.42	3.26
ROUGE-L	27.43	19.32

4.2   Abstractive Summarization

We evaluate the three models on the test dataset and present the results in Table 5. Results on the entire MArSum dataset are also presented in Table 6 for the Pre-trained T5 model used in transfer learning in order to compare to extractive results.

Table 5. Abstractive summarization ROUGE F1 scores obtained on test dataset.

	From-scratch Transformer	Pre-trained T5	Fine-tuned T5
ROUGE-1	5.74	11.50	24.05
ROUGE-2	2.05	3.0	10.16
ROUGE-L	5.57	11.0	23.02

Table 6. Abstractive summarization ROUGE F1 scores obtained on MArSum dataset with the Pre-trained T5 model.

	Pre-trained T5
ROUGE-1	11.50
ROUGE-2	3.0
ROUGE-L	11.0

5   Discussion

It can be seen that the extractive approach has an advantage here, in line with findings often discussed in literature. That said, close scores were obtained with our transfer learning method for abstractive summarization. The extractive approach outperforms the abstractive approach by 5 points for ROUGE-1 and 4.4 points for ROUGE-L. The abstractive approach further achieves better results for ROUGE-2, outperforming the extractive approach by 4.7 points. Even more interesting is that with a relatively small volume of data, we managed to more than double the score of the abstractive approach using transfer learning.

This work being the first to treat ATS for Darija makes these results difficult to compare. This difficulty is illustrated on two levels, firstly the comparison with the MSA results is not very significant since the two variants differ on several levels, notably morphologically and syntactically. Secondly, our dataset is far below the size of the datasets used to obtain the best results in MSA. That said, we present in Table 7 some results obtained for English and MSA.

Table 7. Best ROUGE F1 scores for some English and Arabic (MSA) datasets

	ROUGE-1	ROUGE-2	ROUGE-L
English			
CNN/Daily Mail	44.70 [41]	21.58 [42]	41.60 [43]
XSUM	47.21 [44]	24.56 [44]	25.75 [29]
Arabic (MSA)			
AHS	51.49 [20]	12.27 [20]	51.49 [20]
AMN	44.28 [20]	18.35 [20]	32.46 [20]

In the absence, therefore, of comparable results for Darija, we propose to compare our results with a study presenting a similar scenario to ours, namely a dialect/language considered as low resource with no standardized writing rules and a dataset of similar size to MArSum. Amr et al. [45] worked on Amharic, a low ressource African language, and evaluated their ATS system on a dataset of size 19,000 articles. It turns out that their results are very close to ours (see Table 8) with ROUGE-1 = 20.51, ROUGE-2 = 8.59 and ROUGE-L = 14.76.

Table 8. Comparison with ATS on Amharic language

	Darija (our best model)	Amharic [45]
ROUGE-1	24.05	20.51
ROUGE-2	10.16	8.59
ROUGE-L	23.02	14.76

In the examples in Fig. 5 of the Appendix, we see that even if the from-scratch model fails to produce adequate summaries, it still manages to generate summaries with factual meaning but on different subject. This leads us to deduce that this model could present better results with more training time and more data. Figure 6 of the Appendix shows that the Pre-trained T5 model suffers from the word repetition problem (Example 1) and can generate summaries that do not agree with the source (Example 3) even though it deals with the same subject, these two problems are not found for the Fine-tuned T5 model. Furthermore, we can see from the second example that Fine-tuned model's generated summary is more accurate and concise than the Pre-trained one even if the ROUGE scores are close. Indeed the Pre-trained summary contains some odd words (Highlighted in yellow) with no connection with the original summary.

One of the main challenges when dealing with Arabic dialects in general is the lack of standard orthographies, so the same word can be written in different ways as illustrated in the examples of Table 9, in addition to the fact that the use of prefixes and suffixes does not generally change the meaning. These specificities of the written dialect require models to learn a large vocabulary which impacts the learning time and quality. We can therefore imagine to remedy this problem by using stemming methods and also unify the writing of some words to reduce the size of the vocabulary and improve the learning model. This problem is illustrated in Fig. 7, whereas ROUGE score for these examples is 0, it appears that the model manages to generate some existing words in the article (highlighted in yellow), that said as the word was written in a different way in the article, this word is not counted for the ROUGE score. In the first example, it is the word أساتذة and اساتدة (teachers) written in the article with the 'hamza' ء and without it in the generated summary, whereas in the second example, the word فاجآت (surprised) is wrtten with the suffix و in the generated summary.

Table 9. Examples of different Darija spellings

Darija spellings	MSA	English translation
التيليفيزيون التلفزيون التلفازة التلفزة	التلفاز	Television
موشكيلا موشكة موشكيلة مشكة موشكيل	مشكلة	Problem

We can also generalize the outputs of the model by modifying the titles of the articles so that they are no longer specific to news articles. Indeed, we notice that the model generates titles starting with گود (meaning "Goud"-the newspaper- says:") and also titles ending with the words صور Photos and فيديو Video. Omitting these words from the original headlines may make the model more general for other types of text.

Based on these results, we posit that there is still room for improvement in this model, particularly with regard to the parameters related to the maximum length of the source text and the generated text, which are currently fixed at 128 and 50, respectively, due to limitations related to computing resources. We can also collect more articles and thus increase the size of the training dataset which will improve the performance of the model. In addition, applying more preprocessing steps on the text could also be a source of improvement, for example we can apply stemming methods appropriate to Darija, standardize and generalize article titles by omitting for example the reference to the website name.

6   Conclusion

In this paper, we introduced the first work on Automatic Text Summarization (ATS) for the Moroccan Arabic dialect. Both extractive and abstractive approaches were applied to MArSum, the first dataset dedicated to ATS on Moroccan dialect. We showed that transfer learning based on Transformer models is an efficient solution to this problem, and can achieve state-of-the-art results. Our approach being model-centric, we are looking forward to applying a more data-centric approach by experimenting more preprocessing steps and also gathering more data to improve the summarization results. This paper presents another step for future research for Arabic summarization in general and an important first step for Arabic dialects. It is from this perspective that we release for public use, MArSum dataset and our fine-tuned T5 model for ATS on Moroccan dialect.

A   Appendices

Source Summary	From-scratch Transformer
حنا ما عندناش مع بورطة ولكن باراكا ما ضحكو علينا. راه قرار ما يدخلوش لمغاربة ما عمرو كان بيدو ودابا باش الشوهة كبرات ما خاصوش يتحمل بوحدو المسؤولية هادي قرارات سيادية **Subject:** The management of the problem of citizens stranded abroad because of Covid by the Minister of Foreign Affairs	علاش ما عندناش 5 جي فالمغرب "اتصالات المغرب" واجدة وها فين الموشكيل **Subject:** The reason why 5G network is not available in Morocco
الامير هاري ومراتو ميگان مابقاوش باغيين يعيشو فبريطانيا..وها البلاد اللي اختارو **Subject** : The Prince Harry and his wife are changing country of residence	صحيفة تركية الميريكان عندهم تسجيل صوتي للأمير بن سلمان كايعطي أمر قتل خاشقجي **Subject:** A Turkish newspaper says USA have a recording in which Prince Salman gives orders to kill Khashogji

Fig. 5. Examples of from-scratch model summarization

Source (S), Pre-trained T5 (PT5) and Fine-tuned T5 (FT5) summaries	ROUGE F1 scores
S : واش هما ماشي من العاملين فالصفوف الأمامية؟ القضاة كاعيين على إقصائهم من الفاكسان ضد كورونا **PT5 :** بدات الحملة الوطنية للتلقيح ضد فيروس كوفيد-19. « و » اذكر « إذ » بدل من إذ قبله « أعطيت » بالياء والتاء « الانطلاقة » بالياء والتاء « ديالها الملك محمد السادس ». « و » اذكر « إذ » بدل من إذ قبله « كانوا » أي كفار مكة « ينتظرون » ينتظرون « من » زائدة « من »	[9.3,0,9.3]
FT5 : قضاة كاعيين على إقصائهم من الفاكسان: واش حنا ماشي من العاملين فالصفوف الأمامية؟	[69.23,53.85,46.15]
S : السياحة المغربية: واش بوم-بوم ديال ريدوان غادي ترد المغرب أجمل بلد فالعالم؟ **PT5 :** أغنية جديدة من قبل المنتج العالمي ريدوان. « و » اذكر « إذ » بدل من إذ قبله « وصلت » وصلت « لكثر من 3 ديال المليون مشاهدة ». يريد المنتج العالمي أن يدير الإشهار للثقافة المغربية وللمؤهلات السياحية ديال البلاد.	[9.76,0,9.76]
FT5 : ريدوان" خرج اغنية جديدة على المغرب.. وها علاش"	[10,0,10]
S : بعدما خرجات من الحبس.. سيليا الزياني عايشة الحياة – تصاور **PT5 :** كانت معتقلة سابقة في حراك الريف (سيليا الزياني) بدأت كتخدم في أعمال تلفزية ريفية على القناة الأمازيغية، وفي المسرحيات.	[0,0,0]
FT5 : سيليا الزياني مع ولدها.. عايشة الحياة على راسها – تصاور	[60,33.33, 60]

Fig. 6. Examples of comparison between Pre-trained and Fine-tuned T5 models

Article :

كود الرباط// عاينت "كود" قبل قايا استنفارا كبيرا لدى جميع الاجهزة الامنية من أجل الحسم في فض اعتصام أساتذة التعاقد من عدمه. وعاينت "كود" اجتماعا ميدانيا لقادة المصالح الأمنية على مستوى العاصمة ليلا، بالقرب من النافورة القريبة من محطة القطار. وعاينت "كود" استعدادات مكثقة من اجل التدخل لفض الاعتصام.

S :

عاجل..بالصور.. استنفار في الرباط: استعدادات أمنية لفض اعتصام أساتذة الكونطرا

PT5 :

أساتذة التعاقد باغيين يرجعو للمغرب

Article :

البارح كان عيد ميلاد ديال الممثل التركي مراد يلدريم لي مشهور عند المغاربة بشخصية "أمير" في فيلم "عاصي"، ومراتو المغربية ايمان الباني فاجأتو وسط تقديم حلقة برنامج ديالو. يلدريم كان كيصور برنامج "من سيربح المليون"، باللغة التركية قبل ما تدخل هي والحلوة، والجمهور بدا كيصفق. ونشرت إيمان، فيديو من الاحتفال وكتبات عليها: "عيد ميلاد سعيد.. حبيبي و نور عيني.. وبسمة حياتي.. كل عام وانت الأروع و الأنجح.. كل عام وأنت الفرحة التي تسكن قلبي والشمعة التي تضيء ليالي".

S :

المغربية إيمان الباني فاجآت راجلها التركي في عيد ميلادو — صور وفيديو

PT5 :

"مراد يلدريم ومراتو فاجآتو وسط برنامج "من سيربح المليون

Fig. 7. Generated summaries of poor quality

Title

زينب بنيس كتحارب ملل الڭلاس فالدار بسبب كورونا بطريقتها الخاصة ــفيديو

Text

البارح الخميس، حطات الفنانة التشكيلية زينب بنيس فيديو صوراتو باش تبين فيه كيفاش كتدوز نهارها في الدار مع الحجر الصحي لي مفروض على المغاربة بسباب فيروس كورونا، ولي غادي يستمر الى 20 أبريل الجاي. كيف العادة، بنت نوال متوكل بينات شخصيتها كما هي، وزعزعات مشاعر بعض المحرومين جنسيا بعدما بانت في الأول لابسة بينوار، وشوية لبسات شوميز مشبكة في اللون الأحمر، وباش ماتكترش عليها الهضرة كيف ديما، لبسات معاه سروال ديال الدجين. وبينات زينب أن نهارها كيدوز في الماكلة وفي التكسال، ومن بلاصة لبلاصة حتى دوز هاد المرحلة الحساسة لي عايشها المغرب والعالم كامل.

Title

لونسا كشفات حقيقة بطيخ زاڭورة المعدل جينيا

Text

نفى المكتب الوطني للسلامة الصحية الأخبار اللي كتروج بخصوص زراعة البطيخ المعدل جينيا بمنطقة زاكورة، وكشف أن هذ البطيخ غير معدل جينيا، بحيث المكتب كيشترط لاستيراد جميع الأصناف النباتية الحصول على ترخيص مسبق منه. وزاد المكتب، فبلاغ ليه اطلعات عليه "كود"، أنه كيلزم على مستوردي الأصناف النباتية التوفر على شهادة صادرة على مستنبط الصنف النباتي فالبلد المصدر، كتثبت أن الصنف النباتي المستورد غير معدل جينيا. وكيخص هذ الصنف يكون مسجل فالسجل الرسمي للاصناف النباتية بالمغرب بعد استيفائه لجميع الشروط، منها شهادة اثبات المستنبط ان الصنف غير معدل جينيا بالإضافة إلى إخضاع الصنف المعني لتجارب من طرف مصالح "اونسا" تثبت مطابقته ومردوديته وملائمته لظروف المناخ والتربة.

Title

جواد بينكو بدل أميمة باعزية 360 درجة ــصور

Text

المغنية ديال الشعبي أميمة باعزية بانت بلوك مختلف جدا ومبدلة 360 درجة بعد الماكياج لي دارو ليها جواد بينكو. تصاور مولات "عطيه" دارو حالة في انستڭرام من البارح، وكين لي سولها كاع واش دارت عمليات التجميل حيث تسيفات ديال بصح. هاد التغيير دارتو على قبل شوتينڭ ديال القفطان في رياض في مراكش

Title

ورززات سجلات أكبر عدد دالإصابات بكورونا اليوم والوفيات اللي علنو عليهم اليوم كاملين تسجلو فمراكش.. وها تفاصيل خرى دالإصابات فجميع جهات وأقاليم المغرب

Text

علنات المديريات الجهوية للصحة فجميع الجهات فالمغرب، باستثناء جهة كازا سطات، على حالات الإصابة بفيروس كورونا المستجد فالأقاليم ديالها فهذ 24 ساعة الأخيرة، يعني من الستة ديال لعشية دلبارح الإثنين حتى للستة دالعشية داليوم الثلاثاء. اليوم سجلات جهة درعة تافيلالت أعلى عدد ديال الإصابات، بحيث سجلات 69 إصابة جديدة، وهذ الرقم المرتفع راجع لبؤرة الحبس ديال ورززات اللي سجلات 66 إصابة، حسب بلاغ المديرية العامة لإدارة السجون وإعادة الإدماج، باش يوصل عدد الإصابات فالطوطال فالجهة ل155. كذلك سجلات اليوم جهة طنجة تطوان الحسيمة عدد كبير دالإصابات، اللي وصل ل43 إصابة، وهذشي راجع، حسب ما قال مدير مديرية الاوبئة والوقاية من الأمراض، فندوة اليوم مع الستة، لمجموعة من البؤر فالجهة، حيث تسجلات 14 إصابة جديدة فجوج بؤر صناعية فطنجة (وصلات البؤر ل94 فالمجموع)، بالإضافة لتسجيل 14 إصابة جديدة فبؤرة صناعية فالعرايش (62 فالمجموع)، باش يوصل العدد الإجمالي للإصابات فالجهة ل426 إصابة. وبالإضافة لهذشي تكلم اليوبي اليوم على إصابات جديدة فبؤرة فكازا، وهذشي اللي خلا العدد الإجمالي دالإصابات فيها يطلع من 840 حتى ل855...

Fig. 8. MArSum dataset sample

References

1. Al Qassem, L.M., Wang, D., Al Mahmoud, Z., Barada, H., Al-Rubaie, A., Almoosa, N.I.: Automatic Arabic summarization: a survey of methodologies and systems. PCS **117**, 10–18 (2017). Arabic Computational Linguistics
2. Zaidan, O., Callison-Burch, C.: Arabic dialect identification. Comput. Linguist. **40**(1), 171–202 (2013)
3. Luhn, H.P.: The automatic creation of literature abstracts. JRD **2**(2), 159–165 (1958)
4. Gupta, V., Lehal, G.: A survey of text summarization extractive techniques. JETWI **2**(3), 258–268 (2010)
5. Afsharizadeh, M., et al.: Query-oriented text summarization using sentence extraction technique. In: 2018 4th ICWR, pp. 128–132 (2018)
6. Nagwani, N.: Summarizing large text collection using topic modeling and clustering based on mapreduce framework. JBD **2** (2015). Article number: 6. https://doi.org/10.1186/s40537-015-0020-5
7. Gialitsis, N., et al.: A topic-based sentence representation for extractive text summarization, September 2019
8. Shirwandkar, N.S., et al.: Extractive text summarization using deep learning. In: ICCUBEA 2018, pp. 1–5 (2018)
9. Gupta, S., Gupta, S.K.: Abstractive summarization: an overview of the SOTA. ESA **121**, 49–65 (2019)
10. Nallapati, R., et al.: Abstractive text summarization using seq2seq RNNs and beyond. In: The 20th SIGNLL, Berlin, Germany, pp. 280–290. Association for Computational Linguistics, August 2016
11. Hochreiter, S., Schmidhuber, J.: Long short-term memory. NC **9**(8), 1735–1780 (1997)
12. Song, S., et al.: Abstractive text summarization using LSTM-CNN based deep learning. MTA **78**(1), 857–875 (2019). https://doi.org/10.1007/s11042-018-5749-3
13. Vaswani, A., et al.: Attention is all you need. In: Proceedings of the 31st NIPS, NIPS 2017, Red Hook, NY, USA, pp. 6000–6010. Curran Associates Inc. (2017)
14. Iwasaki, Y., et al.: Japanese abstractive text summarization using BERT. In: 2019 TAAI, pp. 1–5 (2019)
15. Liu, Y., et al.: Text summarization with pretrained encoders. ArXiv, abs/1908.08345 (2019)
16. See, A., et al.: Get to the point: summarization with pointer-generator networks, pp. 1073–1083, January 2017
17. Haboush, A., et al.: Arabic text summerization model using clustering techniques. World Comput. Sci. Inf. Technol. J. **2**(3), 62–67 (2012)
18. Alami, N., Meknassi, M., Noureddine, R.: Automatic texts summarization: current state of the art. J. Asian Sci. Res. **5**, 1–15 (2015)
19. Douzidia, F.S., et al.: Lakhas, an Arabic summarization system (2004)
20. Wazery, et al.: Abstractive Arabic text summarization based on deep learning. Comput. Intell. Neurosci. **2022** (2022). Article ID: 1566890
21. Suleiman, D., Awajan, A.: Deep learning based abstractive Arabic text summarization using two layers encoder and one layer decoder. J. Theor. Appl. Inf. Technol. **98**, 3233 (2020)

22. Al-Maleh, M., Desouki, S.: Arabic text summarization using deep learning approach. J. Big Data **7**(1) (2020). Article number: 109. https://doi.org/10.1186/s40537-020-00386-7
23. Peters, M.E., Zettlemoyer, L., et al.: Deep contextualized word representations. In: Proceedings of the 2018 Conference of the North American Chapter of the ACL: Human Language Technologies, Volume 1 (Long Papers), New Orleans, Louisiana, pp. 2227–2237. ACL, June 2018
24. LeCun, Y., Kavukcuoglu, K., Farabet, C.: Convolutional networks and applications in vision. In: Proceedings of 2010 IEEE ISCS, pp. 253–256 (2010)
25. Devlin, J., et al.: BERT: pre-training of deep bidirectional transformers for language understanding. ArXiv, abs/1810.04805 (2019)
26. Gaanoun, K., Naira, A., Benelallam, I., Allak, A.: DarijaBERT (2021). https://github.com/AIOXLABS/DBert. Accessed 19 Jan 2022
27. Pappagari, R., et al.: Hierarchical transformers for long document classification, pp. 838–844, December 2019
28. Sandhaus, E.: The New York Times Annotated Corpus (2008)
29. Narayan, S., Cohen, S.B., Lapata, M.: Don't give me the details, just the summary! Topic-aware convolutional neural networks for extreme summarization. In: Proceedings of the 2018 Conference on Empirical Methods in Natural Language Processing, Brussels, Belgium, pp. 1797–1807. Association for Computational Linguistics, October–November 2018
30. EL-Haj, M., et al.: Using mechanical Turk to create a corpus of Arabic summaries, January 2010
31. Helmy, M., et al.: Applying deep learning for Arabic keyphrase extraction. Procedia Comput. Sci. **142**, 254–261 (2018). Arabic Computational Linguistics
32. Zaki, A.M., Khalil, M.I., Abbas, H.M.: Deep architectures for abstractive text summarization in multiple languages. In: 2019 14th ICCES, pp. 22–27. IEEE (2019)
33. Miller, D.: Leveraging BERT for extractive text summarization on lectures. ArXiv, abs/1906.04165 (2019)
34. Sutskever, I., et al.: Sequence to sequence learning with neural networks. In: NIPS (2014)
35. Soliman, A.B.: Summarization-Arabic-English-news (2021). https://huggingface.co/marefa-nlp/summarization-arabic-english-news. Accessed 19 Jan 2022
36. Marefa NLP. Arabic T5 Large Model (2021). https://huggingface.co/bakrianoo/t5-arabic-large. Accessed 19 Jan 2022
37. Raffel, C., et al.: Exploring the limits of transfer learning with a unified text-to-text transformer. ArXiv, abs/1910.10683 (2020)
38. Wolf, T., et al.: Transformers: state-of-the-art natural language processing. In: Proceedings of the 2020 Conference on Empirical Methods in Natural Language Processing: System Demonstrations, pp. 38–45. Association for Computational Linguistics, October 2020
39. Steinberger, J., Jezek, K.: Evaluation measures for text summarization. Comput. Inform. **28**, 251–275 (2009)
40. Lin, C.-Y.: ROUGE: a package for automatic evaluation of summaries. In: Text Summarization Branches Out, Barcelona, Spain, pp. 74–81. Association for Computational Linguistics, July 2004
41. Du, Z., et al.: All NLP tasks are generation tasks: a general pretraining framework. CoRR, abs/2103.10360 (2021)
42. Liang, X., et al.: R-Drop: regularized dropout for neural networks. CoRR, abs/2106.14448 (2021)

43. Xiao, D., et al.: ERNIE-GEN: an enhanced multiflow pretraining and fine-tuning framework for natural language generation, pp. 3969–3975, July 2020
44. Zhang, J., et al.: PEGASUS: pretraining with extracted gap-sentences for abstractive summarization, December 2019
45. Zaki, A.M., et al.: AMHARIC abstractive text summarization. CoRR, abs/2003.13721 (2020)

Automatic Change Detection Based on the Independent Component Analysis and Fuzzy C-Means Methods

Abdelkrim Maarir$^{(\boxtimes)}$, Es-said Azougaghe , and Belaid Bouikhalene

Laboratory of Innovation in Mathematics and Applications and Information Technologies (LIMATI), Sultan Moulay Slimane University, Beny Mellal, Morocco
a.maarir@ya.ru

Abstract. Change analysis, an automated process to measure the change on the Earth surface by jointly analyzing two temporally separated images, becomes a significant research domain to understand the changes in land-cover, it provides important knowledge and data to be used in many other fields such as land-cover analyses, mapping generation, planning traffics, etc. This paper describes and evaluates an unsupervised method for change detection in satellite images by following two major steps: The first step focuses on data reduction using the ICA algorithm to improve the efficiency of the classifier. The second step deals with the Fuzzy C-Means classification method to find specified clusters. Changed and unchanged areas are mapped in a binary image. Three different datasets are used to evaluate the result performance of the proposed system, and experiments results show that the used approach can detect changes in multi-temporal satellite images with good accuracy. To show the effectiveness, the comparisons with some other methods from state-of-the-art are shown on multitemporal images captured by Radarsat1 satellite SAR on the Ottawa area.

Keywords: Change detection · Satellite images · ICA algorithm · Fuzzy C-Means · Segmentation

1 Introduction

Spatial remote sensing techniques supply large useful information for an environmental changes analysis, especially, land cover and land use classification [1], mapping generation, planning traffics, etc.

Change detection in images is one of the most applied areas in remote sensing. This issue is becoming an important question in computer vision and the most interesting topics in satellites image processing. Change detection is made between two images taken in two different dates in order to extract things that have been changed in land cover and land use. Based on the keywords related to the change extraction in remote sensing images, the publication records in the two last decades have increased. The statistics indicate that identification of changes in images has become an important area of research. The simplest used methods for change detection are based on algebraic analysis

© Springer Nature Switzerland AG 2022
M. Fakir et al. (Eds.): CBI 2022, LNBIP 449, pp. 178–187, 2022.
https://doi.org/10.1007/978-3-031-06458-6_14

such as differencing and rationing [2, 3] between tow images captured in different times, in addition to the algebraic difference, the Change Vector Analysis (CVA) method, that converts the difference of pixel values into the difference of feature vectors. CVA provides the maximum amount of information on the overall magnitude of changes and the direction of changes between multispectral bands.

Supervised and unsupervised mathematical modelling methods [4–6] can be summarized in the following way: as suggested by [7, 8] changes are extracted bay using feature space transformation. The principal objective is to reduce redundant information using dimensionality reduction by converting the input images into analyzable feature space with principal component analysis (PCA) [9] and hierarchical dimensionality reduction [10].

Change detection of the water body is proposed by [11] using time-series images, this approach is based on the use of Support Vector Machines (SVM) classification method with several extraction indices such as: Automatic Water Extraction Index (AWEI), Normalized Difference Built-up Index (ND-BI), Normalized Water Difference Index (NDWI) and Morphological Building Index (MBI) [12].

Changes are extracted base on buildings detection from satellite images are proposed in several pieces of research [13, 14] by fusing the geographic characteristics related to buildings entity with RS Images [15–17]. Due to its appearance as a group, the distance adjustment between classes are relationship learning (RL) are used by authors [18] to enhance the ability to recognize changing features In addition, other information is used, such as urban light information that shows the density of urban buildings and its details [19], to identify the changes in night lights. Authors [20] suggest that fuzzy clustering methods, such as Fuzzy C-Means, Gustafson–Kessel are more apt to extract changes areas.

Texture-based change detection was proposed by authors in [21], they used the grey-level co-occurrence matrix to extract the texture features and the PCA method for changes classification. In Summary, the above methods of change detection and others are listed in Table 1.

The main purpose of this paper is to extract changes in remote sensing images by using an unsupervised approach. The difference image between T_1 and T_2 is generated by the log ratio operator, then Independent Component Analysis algorithm (ICA) is used to reduce the dimension of the input image, and finally changed and unchanged pixels are identified in two classes by Fuzzy C-Means method. The rest of this paper is structured in the following manner. In Sect. 2, materials and methodologies will be discussed. Section 3 shows the experimental results and discussions. The conclusion and perspectives are reported in Sect. 4.

Table 1. Summary of previous related works on change detection

Methods	Authors	Domains of application	Used method
Algebraic methods	[3]	Land cover	Image differencing
	[2]	Forest area	Change vector analysis
	[11]	Water body	Normalized difference water index
Feature space transformation, classification, and clustering Methods	[22]	Land cover	Markov random field and Spatial information
	[23]		Gabor wavelet features and Fuzzy c-means
	[20]		Gustafson–Kessel and fuzzy c-means
	[1]		Geographic information system analyses
	[8]	Water body	Hierarchical dimensionality reduction Based fuzzy c-means
	[24]	Land cover	Iterative Slow Feature Analysis (ISFA) and Bayesian soft fusion
	[25]		Stationary wavelet transforms, inverse stationary wavelet transform and fuzzy c-means clustering
	[26]		Selective threshold on image histogram and robust semi-supervised fuzzy c-means classifier
	[27]	Land use classification	Support vector machine and maximum likelihood algorithms

2   Materials and Methods

In this section, we provide a brief theoretical background of the ICA and FCM methods for change detection and the used data for testing the proposed approach. We started by presenting the Independent Component Analysis technic used on image processing [28] after difference image calculating, then a FCM classification method is applied in order to separate changed areas from unchanged ones.

2.1   The Used Data

For the considered application, the first used dataset images are acquired by the Landsat LC7/LC8 sensor (with resolution pixels of 30 m × 30 m) in Bine El Ouidane Lake near Beni Mellal city, the dates of acquisition of the images are between 2011 and 2014 as shown in Fig. 1. The second and third datasets are the Ottawa dataset captured by Radarsat-1 satellite SAR sensor in July 1997 and August 1997, the image size is 290 × 350 pixels, line 1 of Fig. 2, and the Yellow River dataset acquired by Radarsat-2 satellite at the Yellow River Estuary in China between 2008 and 2009, the size of image is 257 × 289 pixels, as shown in line 2 of Fig. 2.

2.2 System Specification

The experiments were conducted on an HP computer, Intel Core™ i5-5200U CPU @ 2.20 GHz, 8 GB Memory.

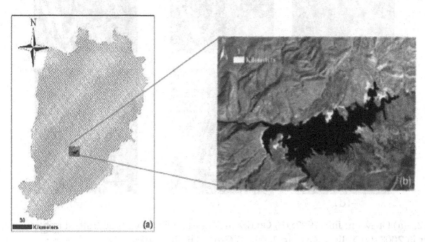

Fig. 1. Study area: (left) Localization of the region of Beni Mellal-Khenifra and (right) Bine El Ouidane region.

2.3 Difference Image and Pre-processing

We use the log-ration operator technique to produce D_I from two identical images at two different times. The difference image (D_I) is generated between the I_1 and I_2 two images acquired at two different times as expressed in Eq. (1).

$$D_I = |\log(I_1 - I_2)| \tag{1}$$

In a pre-processing step, we used a Wiener filter [29] which allows us to remove the noise in the initial images, in order to get a good result and to reduce the processing time. Wiener filter was proposed by N. Wiener [30], it is recognized as Minimum Mean Square Error Filter. It combines both the statistical features of noise and the degradation function into restoration process. The wiener filtering removes the additive noise and inverts the blurring.

2.4 Independent Component Analysis

In a large-scale data classification application, data reduction is a significant technique for data preprocessing. Its goal is to improve the effectiveness and the efficiency of the methods used for classifications in a post processing steps. After difference image generating, the high-dimensional data of D_I are projected to a lower-dimensional space by using the Independent Component Analysis (ICA) method [31] to improve the efficiency of the FCM classifier and the performance of pixel classification.

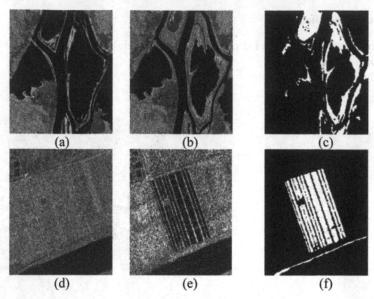

Fig. 2. (a) Ottawa in July 1997; (b) Ottawa in August 1997; (c) Ground truth of (a,b); (d) Yellow River in 2008; (e) Yellow River in 2009; (f) Ground truth of (d,e).

Independent component analysis (ICA)) [32] is based on computing the activation function φ as showing in Eq. (2) by using a computational technique for extracting hidden factors that underlie sets of random variables.

$$\varphi = 0.5 \times y^5 + \frac{2}{3} \times y^7 + \frac{15}{2} \times y^9 + \frac{2}{15} \times y^{11} - \frac{112}{3} \times y^{13} + 128 \times y^{15} - \frac{512}{3} \times y^{17}$$

(2)

where y is the n × m non-negative data matrix D_I, and it based on repeated instructions in order to lead to the following algorithm:

```
The steps of the ICA algorithm:
     Input: [Input features, Output dimension]
     Output: [New features, Reshaped data]
1 Begin
2 Initialization: Move data to the range of [-1,1] after
  whitening to zero mean and unit covariance,
3 Compute the activation function (Eq. 2) in order to
  find the weight matrix,
4 Calculate new features using the weight matrix.
5 End
```

In our case, given an n × m non-negative data matrix D_I that contain the output of the difference image. Significant features and eigenvectors are generated by ICA. Changed and unchanged pixels are classified using fuzzy clustering in the next section.

2.5 Fuzzy Clustering

By dividing the lower-dimensional generated image from the ICA into two classes, the obtained result of the change detection is accurately identified as binary maps related to changed and unchanged pixels. In this section, we are interested in the data clustering method by which large datasets is arranged into clusters of similar groups of data. To classify the pixel of the results image from the previous step on two classes that represent changed and unchanged pixels we used the fuzzy c-means [20, 33]. Fuzzy c-means is an unsupervised method that aims to separate the image into clusters without any prior knowledge of classes [34]. Classes are generated using an iterative method by minimizing an objective function (J) as in Eq. (3) [35], it provides and creates a fuzzy partition of the image by assigning each pixel a degree of membership (between 0 and 1) to a given class.

$$J = \sum_{i=1}^{n} \sum_{j=1}^{c} \|x_i - a_j\|^2 \tag{3}$$

3 Experiment Results and Discussion

The validation and the analyses of the proposed method are done by using three different datasets described in Sect. 2.1 below.

3.1 Performance Evaluation

The performance of the classified image is analyzed using the statistical measure of correct classification based on computing the Overall Error (OE). Correct Classification (CC) is the proportion of correctly detected changed and unchanged pixels among total examined pixels. Mathematically, it can be expressed in Eqs. (4) and (5):

$$OE = False\,Alarms + Missed\,Alarms \tag{4}$$

$$CC = \frac{1 - OE}{(TN + TP + FN + FP)} \tag{5}$$

With FP represents the unchanged pixels which are falsely mapped as changed pixels, TN (True Positive) represents the number of unchanged pixels that correctly detected, FN (False Negative) is the number of changed pixels which are falsely detected as unchanged pixels and TP (True Positive) represents the number of pixels that are correctly detected as changed pixels. The higher CC value means the used method is of good quality. The value CC of the tested images is shown above in Table 2.

It can be seen from Table 2 that the proposed approach indicates a good performance for change detection in all three datasets; it clearly illustrates that the average change detection accuracy is 95.01, 97.50 and 92.33 for Bine El Ouidane dataset, Ottawa dataset and yellow river dataset respectively. Comparisons were made between the proposed

Table 2. Accuracy of the proposed change detection method in three different datasets.

Dataset	Missed alarms	False alarms	Correct detected
Bine El Ouidane dataset	1107	1407	95.01
Ottawa dataset	1489	649	97.95
Yellow River dataset	2929	2935	92.33

method and the GABORTLC [36], RSFCM [26], RFLICM, MRFFCM [4, 37], and PCA-KM [7] methods in order to evaluate its performance.

We selected the Ottawa dataset for testing and change detection. The used results are collected and calculated from the experimental data of published literatures.

According to the comparison of change detection performance results on the Ottawa dataset indicated in Table 3 and Fig. 3, we can see that the proposed method has a lower error value of 2.1% compared to the other methods of the literature. The GABORTLC method proposed by [36] is better on the rate of false pixels (False Alarms) with 253 pixels among the total pixels of the image, but the number of missed pixels (Missed Alarms) is 2531 pixels which generate an increase in the overall error of 2.74%.

The method based on MRFEFCM gives a better rate on decreasing the missed alarms with a value of 712 pixels, but the pixels detected as false alarms are higher with 1636 pixels which generates an overall error of 2.31%.

Table 3. Performance comparison of change detection on the Ottawa dataset produced by different methods.

Methods	False alarms	Missed alarms	Overall error rate	Overall error rate (%)
GABORTLC	253	2531	2784	2.7429
RSFCM	800	1456	2256	2.2227
RFLICM	560	1771	2331	2.2966
MRFFCM	1636	712	2348	2.3100
PCA-KM	955	1515	2470	2.4300
Proposed method	649	1489	2138	2.1064

The proposed ICA based Fuzzy c-mains method allows to obtain good results, with an overall error equal to 2.1064%. Besides, from this result, it is observable that the quantitative analysis confirms the efficiency and effectiveness of the proposed approach for change detection in remote sensing images.

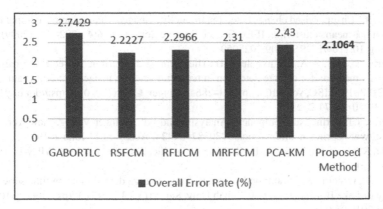

Fig. 3. Average of the overall error rate of detection results on the Ottawa dataset.

4 Conclusion

In this paper, an automatic approach for detecting changes in satellite images by proceeding to dimension reduction based on ICA, followed by a clustering phase using the Fuzzy c-means method has been presented. The log ratio operator was used to produce the difference image between two dates. The significant features obtained by Independent Component Analysis (ICA) algorithm were classified using the fuzzy c-means. The proposed approach is implemented on the images cited above.

The obtained results prove that the proposed approach is good, efficient, reliable, and assumes no prior knowledge about the required scene. Future work will also focus on improving the proposed method to detect change in vegetation and building areas by using object-based techniques.

References

1. Kongapai, P., Sompongchaiyakul, P., Jitpraphai, S.: Assessing coastal land cover changes after the 2004 tsunami using remote sensing and GIS approaches. Walailak J. Sci. Tech. 13(9), 9 (2016)
2. Nackaerts, K., Vaesen, K., Muys, B., Coppin, P.: Comparative performance of a modified change vector analysis in forest change detection. Int. J. Remote Sens. 26(5), 5 (2005). https://doi.org/10.1080/0143116032000160462
3. Lu, D., Mausel, P., Batistella, M., Moran, E.: Land-cover binary change detection methods for use in the moist tropical region of the Amazon: a comparative study. Int. J. Remote Sens. 26(1), 1 (2005). https://doi.org/10.1080/01431160410001720748
4. Gong, M., Zhou, Z., Ma, J.: Change detection in synthetic aperture radar images based on image fusion and fuzzy clustering. IEEE Trans. Image Process. 21(4), 4 (2012). https://doi.org/10.1109/TIP.2011.2170702
5. Hou, B., Wei, Q., Zheng, Y., Wang, S.: Unsupervised change detection in SAR image based on gauss-log ratio image fusion and compressed projection. IEEE J. Sel. Topics Appl. Earth Observ. Remote Sens. 7(8), 8 (2014). https://doi.org/10.1109/JSTARS.2014.2328344
6. Wang, Y., Du, L., Dai, H.: Unsupervised SAR image change detection based on SIFT keypoints and region information. IEEE Geosci. Remote Sens. Lett. 13(7), 7 (2016). https://doi.org/10.1109/LGRS.2016.2554606

7. Celik, T.: Unsupervised change detection in satellite images using principal component analysis and k-means clustering. IEEE Geosci. Remote Sens. Lett. **6**(4), 772–776 (2009). https://doi.org/10.1109/LGRS.2009.2025059

8. Maarir, A., Ider, A.A., Bouikhalene, B.: Hierarchical dimensionality reduction based fuzzy c-means methods for change detection in temporal satellite images. In: Ezziyyani, M. (ed.) AI2SD 2019. AISC, vol. 1105, pp. 273–286. Springer, Cham (2020). https://doi.org/10.1007/978-3-030-36674-2_29

9. Jolliffe, I.T.: Principal Component Analysis. 2nd ed. Springer-Verlag, New York (2002). https://www.springer.com/gp/book/9780387954424. Accessed 11 June 2019

10. Duda, R.O., Hart, P.E., Stork, D.G.: Pattern Classification (2nd Edition). Wiley-Interscience, New York (2000)

11. Sarp, G., Ozcelik, M.: Water body extraction and change detection using time series: a case study of Lake Burdur, Turkey. J. Taibah Univ. Sci. **11**(3), 3 (2017). https://doi.org/10.1016/j.jtusci.2016.04.005

12. Zhu, B., Gao, H., Wang, X., Xu, M., Zhu, X.: Change detection based on the combination of improved SegNet neural network and morphology, pp. 55–59, June 2018. https://doi.org/10.1109/ICIVC.2018.8492747

13. Izadi, M., Saeedi, P.: Automatic building detection in aerial images using a hierarchical feature based image segmentation. In: 2010 20th International Conference on Pattern Recognition, pp. 472–475, August 2010. https://doi.org/10.1109/ICPR.2010.123

14. Maarir, A., Bouikhalene, B., Chajri, Y.: Building detection from satellite images based on curvature scale space method. Walailak J. Sci. Technol. (WJST) **14**(6), 517–525 (2016). 10.14456/vol14iss6pp%p

15. Wang, Q., Zhang, X., Chen, G., Dai, F., Gong, Y., Zhu, K.: Change detection based on Faster R-CNN for high-resolution remote sensing images. Remote Sens. Lett. **9**(10), 923–932 (2018). https://doi.org/10.1080/2150704X.2018.1492172

16. Pirasteh, S., et al.: Developing an algorithm for buildings extraction and determining changes from airborne LiDAR, and comparing with R-CNN method from drone images. Remote. Sens. **11**, 1272 (2019). https://doi.org/10.3390/RS11111272

17. Pang, S., Hu, X., Zhang, M., Cai, Z., Liu, F.: Co-segmentation and superpixel-based graph cuts for building change detection from bi-temporal digital surface models and aerial images. Remote Sens. **11**(6), 6(2019). https://doi.org/10.3390/rs11060729

18. Huo, C., Chen, K., Ding, K., Zhou, Z., Pan, C.: Learning relationship for very high resolution image change detection. IEEE J. Sel. Topics Appl. Earth Observ. Remote Sens. **9**(8), 3384–3394 (2016). https://doi.org/10.1109/JSTARS.2016.2569598

19. Che, M., Gamba, P.: Intra-urban change analysis using sentinel-1 and nighttime light data. IEEE J. Sel. Topics Appl. Earth Observ. Remote Sens. **12**(4), 1134–1142 (2019). https://doi.org/10.1109/JSTARS.2019.2899881

20. Ghosh, A., Mishra, N.S., Ghosh, S.: Fuzzy clustering algorithms for unsupervised change detection in remote sensing images. Inf. Sci. **181**(4), 4 (2011). https://doi.org/10.1016/j.ins.2010.10.016

21. Tomowski, D., Ehlers, M., Klonus, S.: Colour and texture based change detection for urban disaster analysis. In: 2011 Joint Urban Remote Sensing Event, April 2011, pp. 329–332 (2011). https://doi.org/10.1109/JURSE.2011.5764786

22. Gu, W., Lv, Z., Hao, M.: Change detection method for remote sensing images based on an improved Markov random field. Multimed. Tools App. **76**(17), 17719–17734 (2015). https://doi.org/10.1007/s11042-015-2960-3

23. Li, Z., Shi, W., Zhang, H., Hao, M.: Change detection based on Gabor wavelet features for very high resolution remote sensing images. IEEE Geosci. Remote Sens. Lett. **14**(5), 5 (2017). https://doi.org/10.1109/LGRS.2017.2681198

24. Wu, C., Du, B., Cui, X., Zhang, L.: A post-classification change detection method based on iterative slow feature analysis and Bayesian soft fusion. Remote Sens. Environ. **199**(Suppl C), 241–255 (2017). https://doi.org/10.1016/j.rse.2017.07.009

25. Sharma, A., Gulati, T.: Change detection from remotely sensed images based on stationary wavelet transform. Int. J. Elect. Comput. Eng. (IJECE) **7**(6), 6 (2017)

26. Shao, P., Shi, W., He, P., Hao, M., Zhang, X.: Novel approach to unsupervised change detection based on a robust semi-supervised FCM clustering algorithm. Remote Sens. **8**(3), 3 (2016). https://doi.org/10.3390/rs8030264

27. Taati, A., Sarmadian, F., Mousavi, A., Pour, C.T.H., Shahir, A.H.E.: Land use classification using support vector machine and maximum likelihood algorithms by Landsat 5 TM images. Walailak J. Sci. Tech. **12**(8), 8 (2015)

28. Goel, S., Verma, A., Goel, S., Juneja, K.: ICA in image processing: a survey. In: 2015 IEEE International Conference on Computational Intelligence Communication Technology, February 2015, pp. 144–149 (2015). https://doi.org/10.1109/CICT.2015.91

29. Lim, J.S.: Two-dimensional signal and image processing (1990). Accessed 27 Sep 2017. http://adsabs.harvard.edu/abs/1990ph...book.....l

30. Wiener, N.: Extrapolation, Interpolation, and Smoothing of Stationary Time Series: With Engineering Applications, vol. 8. MIT Press, Cambridge (1964)

31. Matteson, D.S., Tsay, R.S.: Independent component analysis via distance covariance. J. Am. Statist. Assoc. **112**(518), 518 (2017). https://doi.org/10.1080/01621459.2016.1150851

32. Shen, H., Jegelka, S., Gretton, A.: Fast kernel-based independent component analysis. IEEE Trans. Signal Process. **57**(9), 9 (2009). https://doi.org/10.1109/TSP.2009.2022857

33. Amin, M.R.M., Bejo, S.K., Ismail, W.I.W., Mashohor, S.: Colour extraction of agarwood images for fuzzy c-means classification. Walailak J. Sci. Technol. (WJST) **9**(4), 445–459 (2012). https://doi.org/10.2004/wjst.v9i4.211

34. Dunn, J.C.: A fuzzy relative of the ISODATA process and its use in detecting compact well-separated clusters. J. Cybern. **3**(3), 3 (1973). https://doi.org/10.1080/01969727308546046

35. Yang, M.-S.: A survey of fuzzy clustering. Math. Comput. Modell. **18**(11), 11 (1993). https://doi.org/10.1016/0895-7177(93)90202-A

36. Li, H.C., Celik, T., Longbotham, N., Emery, W.J.: Gabor feature based unsupervised change detection of multitemporal SAR images based on two-level clustering. IEEE Geosci. Remote Sens. Lett. **12**(12), 12 (2015). https://doi.org/10.1109/LGRS.2015.2484220

37. Gong, M., Su, L., Jia, M., Chen, W.: Fuzzy clustering with a modified MRF energy function for change detection in synthetic aperture radar images. IEEE Trans. Fuzzy Syst. **22**(1), 1 (2014). https://doi.org/10.1109/TFUZZ.2013.2249072

Sentiment Analysis Decision System
for Tracking Climate Change Opinion in Twitter

Mustapha Lydiri[1]([✉]) [iD], Youssef El Habouz[2], and Hicham Zougagh[1] [iD]

[1] TIAD Laboratory, Sciences and Techniques Faculty, Sultan Moulay Slimane University,
PB 523, Beni Mellal, Morocco
Mustapha.lydiri@usms.ma
[2] IGDR, UMR 6290 - CNRS – Rennes 1 University, Rennes, France
youssef.elhabouz@univ-rennes1.fr

Abstract. Global warming or climate change is one of the most trend topics of
the decade in the world, according to scientists the earth is getting warm more
every year, hence people are more and more complaining about this phenomenon,
some of them believe that climate change is happening, and we should worry and
act about it. Despite that, the Intergovernmental Panel on Climate Change (IPCC)
confirms that global warming is real and causes climate change, the majority of
people are still in doubt about that, this group is generally called deniers or skeptics
who think that it is not real and not caused by human. This group of people is
costing a lot for countries, by affecting others who think that we should take
corrective actions toward global warming. Thus, it is required to create models
that identify the impact of people's thoughts to help governments for achieving
better control of their citizens. In this work we present a new model for analyzing
public opinion on social media platforms especially Twitter about climate change
subject, we adopted the Sentiment Analysis technique, which is a field of natural
language processing, we provided an effective model based on Convolutional
Neural Network (CNN) for detecting people's reviews on climate change in social
media platforms. Our model may assist decision-makers in producing appropriate
strategies to ameliorate the impacts of the climate change phenomenon.

Keywords: Twitter · Sentiment Analysis · Climate change

1 Introduction

Climate change has become a disaster that citizens are getting more concerned about, as
a result of rapid global urbanization and gases emission in the word, especially United
States which is the largest source of greenhouse gas emissions from many activities such
as transportation, electricity production, industry, commercial and residential.

Climate change and its implications have become a major subject of concern across
the world. We must lower our carbon footprint in order to maintain life on Earth. Scientists
are urging various organizations and governments to limit their co2 emissions. As a result,
different governments and organizations have cut down public education. Companies

M. Fakir et al. (Eds.): CBI 2022, LNBIP 449, pp. 188–196, 2022.
https://doi.org/10.1007/978-3-031-06458-6_15

are also helping to lead the charge by changing their goods to minimize their carbon impact.

Social media networks are the best source of data on climate change-related subjects for gaining an understanding of public opinion. Twitter is a micro-blogging site that allows its users to post a "tweet" which is a text limited to 140 characters, each tweet can be identified by a "hashtag", which present the subject of the post, this keyword can help others to find all posted tweets of any specific subject. Nowadays Twitter is used by the most famous celebrities, politicians, and the majority of people around the world. All of that makes it a good source of opinion. Text mining or text analytics is a way of analyzing unstructured data to uncover patterns and forecast popular attitudes, which may help businesses to make better decisions.

Sentiment analysis is a field of natural language processing (NLP), this technique is usually applied to text for extracting the real sentiment expressed by authors toward subjects such as products, services, events,... It could be done by using lexicon-based approaches with the use of a dictionary of words and sentiment class polarity or via text classification methods with the use of machine learning algorithms to build text classification models. Sentiment analysis includes many methods such as Document-Level which means to assign a sentiment class to a textual document, or Sentence-Level which explores the sentiment carried in each sentence separately.

In this paper, we aim to establish a model for predicting the sentiment of tweets discussing climate change on Twitter. We will use Twitter data in this research since it is a great source of text data, which is relevant in analyzing and understanding people's views on climate change.

This paper is organized as follows: we first introduce some related work, followed by introducing the Sentiment Analysis approach. Then we describe our approach in Sect. 4 along with the data used. Additionally, the results are given and discussed in Sect. 5, at the end we give a conclusion and future aims.

2   Related Work

Many works are done in the sentiment analysis field in social media platforms especially Twitter microblogging site. [1] proposed a novel Machine learning approach for sentiment analysis on Twitter and IMDB website by combining ULMFiT (Universal Language Model Fine-tuning) with SVM model. The results proved that the model increases detection efficiency and accuracy on the Twitter dataset. [2] tried to analyze Iranian people's view toward COVID-19 vaccination and compare opinions about domestic and imported COVID-19 vaccinations, in order to accomplish this study, they retrieved 803278 tweets mentioning COVID-19 subject in Iran country from Twitter. The research found an unfavorable opinion toward both domestic and imported vaccinations in recent months, also a major decrease of desire to take the vaccine. [3] presented research on development on Rinca Island by the Indonesian government, with analyzing Thousands of people's views shared on social media, particularly Twitter. The study examined the public's perceptions of the development on Rinca Island and sorted it into three categories, the techniques used in this research are Doc2Vec model to transform tweets into vectors and support vector machines and logistic regression algorithms to classify those tweets into tree sentiment classes.

Much research focused on climate change opinion in Twitter, for example [4] combined topic modeling approach and sentiment analysis technique to find out climate change aspect in extracted posts from Twitter using the relevant keywords. [5] used topic modeling to detect users who believe that climate change is anthropogenic. This research used the 'global warming' keyword to search for opinions that believe in human causation or a result of natural cycles in Twitter. The results demonstrate substantial political polarization, as well as major vocabulary discrepancies. [6] presented a review on the way climate change is discussed on Twitter, the study focused on reactions to climate change-related events, and different areas of research about discussions of climate change topic on Twitter. Similar work used deep learning to detect climate change deniers on Twitter [7], by developing an optimized Deep Neural Network (DNN) classifier to examine public attitudes toward global warming, they used special keywords to identify target tweets, then labeled those tweets into denier and non- denier with the use of a lexicon-based method. For the vectorizing part, they used the continuous bag-of-words (CBOW) model for converting tweets into vectors with the same length. The results of the DNN based model were robust, but the research has many limitations such as using a small dataset of tweets, and manual labeling which can be time-consuming.

3 Sentiment Analysis

Sentiment Analysis or opinion mining is field of Natural language processing, generally this technique is performed on textual data, in order to detect and examine opinion or feedback of the author. It is the key factor businesses to find out their customer feedback about any product or service. Sentiment Analysis could be performed via many strategies, one is with the use of lexicon-based approaches, this one rely on dictionaries of words along with the sentiment of each one, thus the final sentiment of the document is generated based on the final sentiment score calculated from words [8]. In other hand text classification approaches, which is based on machine learning classifiers to classify textual data and detect the sentiment carried. Finally hybrid approach, this technique combine text classification lexicon-based and approaches. Many research used sentiment analysis such as [9], they used a hybrid topic based sentiment approach, to detect sentiment polarity of tweets for election prediction. [10] adopted sentiment analysis for crime rate detection in India along with detecting states that have highest number of crimes. In the field of healthcare [11] Designed a module for obtaining sentiments and emotions related to healthcare domain.

4 Proposed Approach

Convolution Neural Network or CNN [12] is a powerful deep learning algorithm for processing image content. For its effectiveness in the image classification field, we have the right to assume that CNN can be used in the field of natural language processing, due to its great performance. One of the CNN advantages is that it can extract only the pertinent features from data, and because of the tweet characteristics it can contain a lot of non-relevant words or symbols related to the sentiment expressed in tweets. Thus, it is required to adopt CNN algorithm for sentiment classification of tweets related to climate change subject. The general architecture of our proposed approach is shown in Fig. 1.

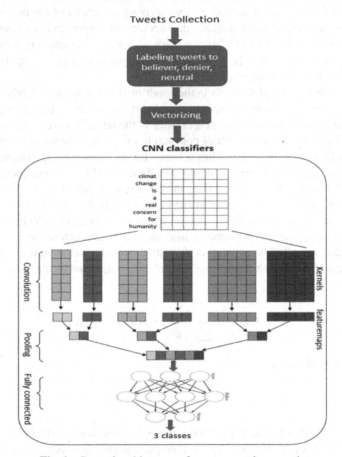

Fig. 1. General architecture of our proposed approach.

Step1: The first step of our model is collecting data. Generally, Twitter data is collected with the use to Twitter Streaming API which can provide full access of Twitter content, and special keywords for targeting climate change related tweets such as (#climatechange, #globalwarming, #environment, #climateaction,..).

Step2: The labeling process of generated tweets to believers and deniers could be done manually, for deciding the tweet if it indicates an opinion that supports climate change action, or against the existence of man-made climate change, but this process is highly time-consuming. In our research, we used the Hashtag keyword mentioned in each tweet to decide the label class of the tweet. For example, tweets that contains (#ClimateHoax, #ClimateChangeIsNotReal) is categorized as deniers, and those that contain (#ClimateChangeIsReal, #ActonClimatChange) is labeled as believers.

Step3: The third step of our approach is vectorizing sentences of the tweets. We used word2vec algorithm to transform words into vectors, this algorithm is a neural network proposed by Google in 2013 [13], it can learn vector representation of words and the results is a sequence of words:$[s_i, .., s_j]$, the embedding matrix $[w_i, .., w_j]$ is formed by concatenating all words sequences.

Step4: The matrix resulted in step3 is the input of the first layer of CNN, the **convolutional** layer which aims to extract good and pertinent features the input data of this layer is a matrix, with the use of kernels or filter **F**, the result is a vector C calculated by: $C_i = (W * F)_i + b_i$. C is the result of computing the product between a column of the matrix W and a filter K and then adding a bias b for each feature map. The result is feed to a non-linear activation function Sigmoid or Tanh, or RELU. In our model, we used RELU which simply take positive values or zero value: max(0, x).

Step5: The next layer of CNN is the **pooling** layer, it can find the feature that corresponds to the filter window, by selecting the maximum value. This layer's job is to collect just the most significant features with the highest value for each feature map.

Step6: A **fully connected** layer receives the output of the final convolutional and pooling layers x. The probability distribution over the labels is calculated by the Softmax function [14] as follow:

$$P_i(x) = \frac{e^{(x_i)}}{\sum_{j=1}^{K} e^{x_j}}, i = 1, 2, \ldots, K \tag{1}$$

K is the number of classes.

4.1 Dataset

The data used in this research is a collection of tweets funded by [15]. The dataset gathers 43943 tweets related to climate change collected between Apr 27, 2015, and Feb 21, 2018. In total. Each tweet is labeled as one of the following classes: Positive (tweets contain sentiment believing in climate change), Negative (tweets deny that climate change is real), Neutral (the tweets neither support nor refute the belief of climate change disaster). The numbers of tweets in each sentiment class are given in Fig. 2. "0" indicates positive tweets, "1" for negative ones and "2" is neutral ones.

Figure 3 present the most frequently used words in our data, as we can deduce from this figure that most of the words are related to the subject of climate change.

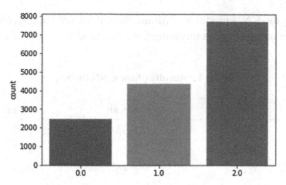

Fig. 2. Numbers of tweets in each sentiment classes

Fig. 3. Most frequently used words in our dataset.

5 Result and Discussion

For evaluating our model based on CNN, we used the following metrics:

$$Precision = \frac{TP}{TP + FP} \tag{2}$$

$$Recall = \frac{TP}{TP + FN} \tag{3}$$

$$F_1 = 2 * \frac{Precision * Recall}{Precision + Recall} \tag{4}$$

$$Accuracy = \frac{TP + TN}{TP + FP + TN + FN} \tag{5}$$

The number of true positives TP and true negatives TN are the number of effectively classified positive and negative tweets respectively. Likewise, false positives FP and false negatives FN refer to the numbers of positive and negative tweets incorrectly classified.

After training the CNN model to climate change related tweets containing people's opinions about the subject of climate change, we obtained the results illustrated in Table 1.

Table 1. Results of the CNN model.

	Precision	Recall	F1-mesure
Positive	0.86	0.95	0.95
Negative	0.73	0.58	0.65
Neutral	0.80	0.86	0.83

The results indicate that our model can classify most of positive, negative, and neutral sentiment tweets. With more detailed results in Fig. 4, the confusion matrix describes the samples successfully classified and the misclassified ones of each class, we note "0" for positive, "1" for negative and "2" for neutral sentiment. We can observe the number of tweets correctly classified (454 of positive, 503 of negative and 1338 of neutral tweets), and incorrectly classified (76 of positive, 188 of negative, and 342 of neutral tweets). The model then achieved to classify 80% of the testing tweets.

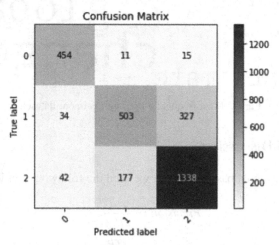

Fig. 4. Confusion matrix

To prove the robustness of our model we compared our results with other similar works in the same field in Fig. 5 in term of accuracy metric.

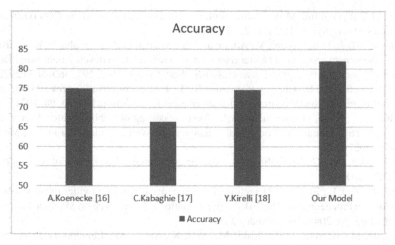

Fig. 5. Related work results compared with our model

With these results above we can deduce that our model is efficient and good for the classification of tweets related to climate change subject. Thus, the CNN based model for detecting climate change opinion could be a powerful system for detecting reviews and opinions of people on Twitter which gathers most of the world population. This work could be extended to examine other domains, also with using other deep leaning algorithms could help to make the model more accurate.

6 Conclusion

In this paper, we developed a new system of understanding people's opinions through social media platforms especially Twitter on the subject of climate change, with the use of sentiment analysis technique. This work present a good tool for deciders, it could help government to make the good decisions, to predict and prevent any disturbance in any country. The major limitation of this work is labeling collected Twitter data which is time consuming if it is done manually, and hard with the use of machines due to the ambiguity of people's posts or opinions and the use of sarcasm makes detection of the real sentiment difficult. This research can be expanded to explore more subjects such as elections [19], reviews on products [20], politics [21], and more fields.

In the future, we plan to explore more techniques for improving the efficacity of our model such as attention mechanism, also using more deep learning algorithms to build a more accurate model for sentiment analysis. Also, we aim to expand the sentiment classes used to include all possible classes.

References

1. AlBadani, B., Shi, R., Dong, J.: A novel machine learning approach for sentiment analysis on Twitter incorporating the universal language model fine-tuning and SVM. Appl. Syst. Innov. **5**(1), 13 (2022)

2. Nezhad, Z.B., Deihimi, M.A.: Twitter sentiment analysis from Iran about COVID 19 vaccine. Diabetes Metab. Syndr. **16**(1), 102367 (2022)
3. Hidayat, T.H.J., Ruldeviyani, Y., Aditama, A.R., Madya, G.R., Nugraha, A.W., Adisaputra, M.W.: Sentiment analysis of twitter data related to Rinca Island development using Doc2Vec and SVM and logistic regression as classifier. Proc. Comput. Sci. **197**, 660–667 (2022)
4. Becken, S., Stantic, B., Chen, J., Connolly, R.M.: Twitter conversations reveal issue salience of aviation in the broader context of climate change. J. Air Transp. Manag. **98**, 102157 (2022)
5. Al-Rawi, A., Kane, O., Bizimana, A.J.: Topic modelling of public Twitter discourses, part bot, part active human user, on climate change and global warming. J. Environ. Med. **2**(1), 31–53 (2021)
6. Fownes, J.R., Yu, C., Margolin, D.B.: Twitter and climate change. Sociol. Compass **12**(6), e12587 (2018)
7. Chen, X., Zou, L., Zhao, B.: Detecting climate change deniers on twitter using a deep neural network. In Proceedings of the 2019 11th International Conference on Machine Learning and Computing, pp. 204–210, February 2019
8. Taboada, M., Brooke, J., Tofiloski, M., Voll, K., Stede, M.: Lexicon-based methods for sentiment analysis. Comput. Linguist. **37**(2), 267–307 (2011)
9. Bansal, B., Srivastava, S.: On predicting elections with hybrid topic based sentiment analysis of tweets. Proc. Comput. Sci. **135**, 346–353 (2018)
10. Khan, R., et al.: Crime Detection Using Sentiment Analysis (2021)
11. Ramirez-Tinoco, F.J., et al.: Use of sentiment analysis techniques in healthcare domain. In: Alor-Hernández, G., Sánchez-Cervantes, J.L., Rodríguez-González, A., Valencia-García, R. (eds.) Current Trends in Semantic Web Technologies: Theory and Practice. Studies in Computational Intelligence, vol. 815, pp. 189–212. Springer, Cham (2019). https://doi.org/10.1007/978-3-030-06149-4_8
12. Lydiri, M., El Mourabit, Y., El Habouz, Y.: A new sentiment analysis system of climate change for smart city governance based on deep learning. In: Ben Ahmed, M., Rakıp Karaş, İ, Santos, D., Sergeyeva, O., Boudhir, A.A. (eds.) Innovations in Smart Cities Applications Volume 4. LNNS, vol. 183, pp. 17–28. Springer, Cham (2021). https://doi.org/10.1007/978-3-030-668 40-2_2
13. Liu, H.: Sentiment analysis of citations using word2vec. arXiv preprint arXiv:1704.00177 (2017)
14. Gao, B., Pavel, L.: On the properties of the softmax function with application in game theory and reinforcement learning. arXiv preprint arXiv:1704.00805 (2017)
15. Canada Foundation for Innovation JELF Grant to Chris Bauch, University of Waterloo
16. Koenecke, A., Feliu-Fabà, J.: Learning twitter user sentiments on climate change with limited labeled data. arXiv preprint arXiv:1904.07342 (2019)
17. Kabaghe, C., Qin, J.: Classifying tweets based on climate change stance. Training **66**(60.9), 61 (2020)
18. Kirelli, Y., Arslankaya, S.: Sentiment analysis of shared tweets on global warming on Twitter with data mining methods: a case study on Turkish language. Comput. Intell. Neurosci. **2020** (2020)
19. Xia, E., Yue, H., Liu, H.: Tweet sentiment analysis of the 2020 US presidential election. In Companion Proceedings of the Web Conference 2021, pp. 367–371, April 2021
20. Yang, L., Li, Y., Wang, J., Sherratt, R.S.: Sentiment analysis for E-commerce product reviews in Chinese based on sentiment lexicon and deep learning. IEEE Access **8**, 23522–23530 (2020)
21. Antypas, D., Preece, A., Collados, J.C.: Politics and Virality in the Time of Twitter: A Large-Scale Cross-Party Sentiment Analysis in Greece, Spain and United Kingdom. arXiv preprint arXiv:2202.00396 (2022)

Analysis of Decision Tree Algorithms for Diabetes Prediction

Youssef Fakir[✉] and Naoum Abdelmotalib

Sultan Moulay Slimane University, Beni-Mellal, Morocco
info.dec07@yahoo.fr

Abstract. Data Mining (DM) is a helpful tool to extract and exploit the information from a large data set. There are different methods and algorithms available in data mining field. Several DM algorithms are used for classification such as Artificial Neural Network (ANN), K-Nearest Neighbor (K-NN), etc. The Decision Tree (DT) mining remains the best algorithm. In this paper, different classification methods including decision tree, C-RT, C5.0, AD-Tree and CS-MC4 algorithms are presented. These algorithms are evaluated using Recall, precision and F-measure. Experimental results show that AD-Tree is faster and present higher accuracy than the other classifier using a Diabetes data set.

Keywords: Data mining technique · Classification techniques · Decision tree · C5.0 · C-RT · AD-Tree · CS-MC4 · Diabetes dataset

1 Introduction

Diabetes is a disease [1–3] due to the insufficient production of insulin in the pancreas. It can lead to serious complications. Patient-age and the glucose-concentration in the blood-plasma are crucial for diabetes prediction. Several methods [4–8] have been used for prediction in the diabetes diagnosis in patient such as DT, K-NN [9], SVM [7, 8] and ANN [10–15]. Decision Tree (DT) [16, 17] is a powerful data mining method and offers the simplest tree structure for the user to make decision. For classifying the data, the user divided it into two sets, one for training and the other for testing.

In this paper, we present the experimental analysis using the well-known DT algorithms such as C5.0 [18, 19], AD-Tree [20], C-RT [21, 22] and CS-MC4 [23–25] on diabetes data set in order to get out a conclusion which algorithm is the best one for prediction. DT generates the rule for classifying the data set [26]. The attribute and split point that best split the training instance belonging to this leaf is the goal at each node. Figure 1 represents a structure of the DT.

This paper is organized as follows: Sect. 2 deals with performance measures used. Section 3 describes experimental results. Conclusion is given in Sect. 4.

© Springer Nature Switzerland AG 2022
M. Fakir et al. (Eds.): CBI 2022, LNBIP 449, pp. 197–205, 2022.
https://doi.org/10.1007/978-3-031-06458-6_16

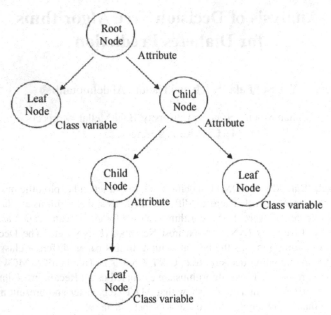

Fig. 1. Decision tree structure

2 Performance Measures

The performance of the algorithms is known by calculating the accuracy, recall, specificity and F-measure which are depicted by using a diabetic dataset and the confusion matrix given in Table 1.

Table 1. Confusion matrix

True-values			
Prediction	Total-population	Positive	Negative
	Positive	T_P	F_P
	Negative	F_N	T_N

T_P is the True-Positive, T_N is the True-Negative, F_N is the False-Negative and F_P is the False-Positive.

The accuracy using confusion matrix is given by Eq. 1:

$$Accuracy = \frac{T\_P + T\_N}{T\_P + F\_P + F\_N + T\_N} \tag{1}$$

We can deduce the Error-Rate (ER) from accuracy using this formula.

$$ER = 1 - Accuracy \tag{2}$$

-Precision is given by:

$$\text{Precision } \frac{T\_P}{T\_P + F\_P} \tag{3}$$

-Specificity is the total number of true negative predictions divided by the total number of negatives.

$$\text{Specificity} = \frac{T\_N}{T\_N + F\_P} \tag{4}$$

Recall is the total number of positive prediction divided by the total number of positives. Recall ad F-measure are given respectively in Eq. (5) and (6).

$$\text{Recall} = \frac{T\_P}{T\_P + F\_N} \tag{5}$$

$$\text{F-measure} = \frac{2 * \text{Re}call * \text{Pr}ecision}{\text{Re}call + \text{Pr}ecision} \tag{6}$$

3 Experimental Results

In this paper, the dataset contains sixteen attributes namely Patient- number-, -Cholesterol-, -Glucose-, -Hdl_chol-, -Chol_hdl-, -Chol_hdl-ratio-, -Age-, -Gender-, -Height-, -Weight-, -BMI-, -Systolic_bp-, -Waist-, -Hip-, Waist_hip_ratio-, -Diabetes- (see Table 2). The data was collected from 390 patients at the NIDDKD (USA). The data file is in an Excel format to be compatible with Weka.

To classify the diabetes disease, the error rates and accuracy are calculated. The confusion matrix (represented in Table 3) is used in order to find the various evolution measures like accuracy, T_P rate, -precision, -recall, etc. The performance of algorithms C5.0, C-RT, AD-Tree and CS-MC4 are given respectively in Tables 5, 6, 7, 8 and 9.

Table 2. Description of data attributes

Variables	Description
-Patient_number-	-Number of the patient
-Cholesterol-	-Type of Lipids
-Glucose-	-Sugar molecules
-Hdl-chol-	-High density lipoprotein
-Chol_hdl_ratio-	-Cholesterol ratio
-Age-	-Age of patient
-Gender-	-Male or female
-Height-	-Height of the patient

(*continued*)

Table 2. (*continued*)

Variables	Description
-Weight-	-Weight of the patient
-BMI-	-Body Mass Index
-Systolic-BP-	-Systolic Blood Pressure
-Waist-	-Waist measurement
-Hip-	-Hip measurement
-Waist_hip-ratio-	-Waist and hip ratio
-Diabetes-	-Diabetes or no diabetes
-Chol_hdl-	-Good cholesterol

The AD-Tree algorithm (Table 3) shows the best results with 17 minimum of false diabetes and 43 maximum of true diabetes, while the other algorithms show less accurate results. Figure 2 illustates the correctly and incorrectly classified intances. Here, AD-Tree has classified correctly the maximum number of instances with value of 356 and shows the lowest score of incorrectly classified instances of 34. Table 4 shows the accuracy of each method and its graphic representation is illustrated in Fig. 3.

Table 3. Confusion matrix of different algorithms

Classifier	Desired results	Prediction	
		No diabetes	Diabetes
C5.0	No diabetes	308	22
	Diabetes	23	37
C-RT	No diabetes	315	15
	Diabetes	35	25
AD Tree	No diabetes	313	17
	Diabetes	17	43
CS-MC4	No diabetes	314	16
	Diabetes	22	38

The accuracy chart of Fig. 3 shows that with its 91,28% the AD Tree algorithm is the most accurate. Tables 5, 6, 7, 8 and 9 and Fig. 4 present the experimental results using TP rate, Precision, Recall and F-measure. Table 10 shows the computing time done by each algorithm.

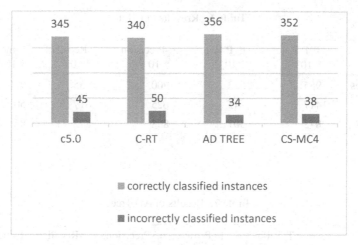

Fig. 2. Instances classification

Table 4. Accuracy of different methods

Classifier	C5.0	C-RT	AD Tree	CS-MC4
Accuracy	88,46%	87,17%	91,28%	90,25%

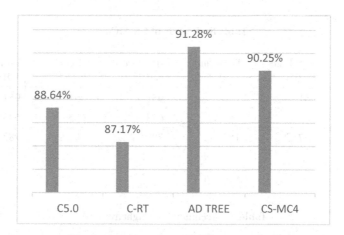

Fig. 3. Accuracy chart of different methods

Table 5. Results of C5.0

Class	T_P-Rate	F_P-Rate	-Precision	-Recall	F-measure
No diabetes	0,933	0,383	0,931	0,933	0,932
diabetes	0,617	0,067	0,627	0,617	0,622
Weight avg	0,885	0,335	0,884	0,885	0,884

Table 6. Rresults of C-RT

Class	T_P-Rate * 10^{-3}	F_P-Rate * 10^{-3}	-Precision * 10^{-3}	-Recall * 10^{-3}	F-measure * 10^{-3}
No diabetes	955	583	900	955	926
Diabetes	417	45	625	417	500
Weight avg	872	501	858	872	861

Table 7. Results of AD Tree

Class	T_P-Rate * 10^{-3}	F_P-Rate * 10^{-3}	-Precision * 10^{-3}	-Recall * 10^{-3}	F-measure * 10^{-3}
No diabetes	948	283	948	948	948
Diabetes	717	52	717	717	717
Weight avg	913	248	913	913	913

Table 8. Results of CS-MC4

Class	T_P-Rate * 10^{-3}	F_P-Rate * 10^{-3}	-Precision * 10^{-3}	-Recall * 10^{-3}	F-measure * 10^{-3}
No diabetes	951	36	951	951	951
Diabetes	703	65	633	633	633
Weight avg	827	183	792	792	782

Table 9. Accuracy by weight average

Parameter	C5.0	C-RT	AD Tree	CS-MC4
T_P-Rate	0,885	0,885	0,913	0,827
F_P-Rate	0,335	0,335	0,248	0,183
-Precision	0,884	0,884	0,913	0,792
-Recall	0,885	0,885	0,913	0,792
F-measure	0,884	0,884	0,913	0,792

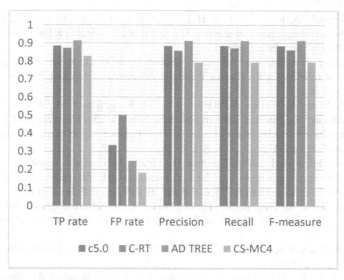

Fig. 4. Weight average of different parameter

Table 10. Computing time of different classifiers

Classifier	C5.0	C-RT	AD Tree	CS-MC4
Time (s)	0.02	0.33	0.06	0.4

4 Conclusion

In this study, four algorithms were used in order to predicte diabete. Different tests are conducted over a chosen dataset of 390 patients with its specific set of attributes. The four algorithms are evalued using various measures such as recall, precision and F-measure. Here the studies conclude that the AD tree classifier acheive higher test accuracy of 91,28% and is faster than the other algorithms.

References

1. Matuszewski, W., et al.: Prevalence of Diabetic Retinopathy in Type 1 and Type 2 Diabetes Mellitus Patients in North-East Poland. Medecina (2020)
2. Roy, M.S., et al.: The prevalence of diabetic retinopathy among adult type 1 diabetic persons in the United States. Arch. Ophthalmol. **122** (2004). (©2004 American Medical Association)
3. Wang, S.Y., Andrews, C.A., Herman, W.H., Gardner, T.W., Stein, J.D.: Incidence and Risk Factors for Developing Diabetic Retinopathy among Youths with Type 1 or Type 2 Diabetes throughout the United States, American society of ophthalmology (2017) https://doi.org/10.1016/j.ophtha.2016.10.031
4. Fiarni, C., Sipayung, E.M., Maemunah, S.: Analysis and prediction of diabetes complication disease using data mining algorithm. In: The Fifth Information Systems International Conference 2019, Science Direct. Procedia Computer Science, vol. 161, pp. 449–457 (2019)

5. Gárate-Escamila, A..K.., Hassani, A..H..E.., Andrès, E..: Classification models for heart disease prediction using feature selection and PCA. Inf. Med. Unlock. **19**, 100330 (2020). https://doi.org/10.1016/j.imu.2020.100330

6. Mujumdar, A., Vaidehi, V.: Diabetes prediction using machine learning algorithms. In: International Conference on Recent Trends in Advanced Computing 2019, ICRTAC 2019 (2019)

7. Ghosh, P., Azam, A., Karim, A., Hassan, M., Roy, K., Jonkman, M.: A comparative study of different machine learning tools in detecting diabetes. 25th International Conference on Knowledge-Based and Intelligent Information & Engineering Systems. Procedia Comput. Sci. **192**, 467–477 (2021)

8. Viloria, A., Herazo-Beltran, Y., Cabrera, D., Pineda, O.B.: Diabetes diagnostic prediction using vector support machines. In: The 11th International Conference on Ambient Systems, Networks and Technologies (ANT), 6–9 April 2020, Warsaw, Poland (2020)

9. Zhang, X., Xiao, H., Gao, R., Zhang, H., Wang, Y.: K-nearest neighbors rule combining prototype selection and local feature weighting for classification. Knowl. Based Syst. **243** (2022)

10. Patel, B.R., Rana, K.K.: A survey on decision tree algorithm for classification. Int. J. Eng. Dev. Res. **2**(1) (2014)

11. Sisodia, D., Sisdia, D.S.: Prediction of diabetes using classification algorithms. In: International Conference on Computational Intelligence and Data Sciences (ICCIDS), Science Direct Procedia Computer Science, vol. 132, pp. 1578–1585 (2018)

12. Harz, H.H., Rafi, A.O., Hijazi, M.O., Abu-Naser, S.S.: Artifical neural network for diabetes using JNN. Int. J. Acad. Eng. Res. **4**(10), 14–22 (2020)

13. Liu, J., Tang, Z.H., Zeng, F., Li, Z., Zhou, L.: Artificial neural network models for prediction of cardiovascular autonomic dysfunction in general Chinese population. BMC Med. Inf. Dec. Mak. **13**(1) (2013). https://doi.org/10.1186/1472-6947-13-80

14. Pradhan, N., Rani, G., Dhaka, V.S., Poonia, R.C.: Diabetes prediction using artificial neural network. Deep Learn. Tech. Biomed. Health Inf. **121**, 327–339 (2020). https://doi.org/10.1016/B978-0-12-819061-6.00014-8

15. Temurtas, H., Yumusak, N., Temurtas, F.: A comparative study on diabetes disease diagnosis using neural networks. Expert Syst. Appl. **36**(4), 8610–8615 (2009). https://doi.org/10.1016/j.eswa.2008.10.032

16. Sharma, A.K., Sahni, S.: A comparative study of classification algorithms for spam email data analysis. Int. J. Comput. Sci. Eng. **3**(5), 1890–1895 (2011)

17. Nemae, D.R., Gupa, R.K.: Diabetes prediction using BPSO and decision tree classifier. In: 2nd International Conference on Data, Engineering and Applications (IDEA), IEEE Xplore 2020 (2020)

18. Nancy, P., Ramani, R.G., Jacob, S.G.: Discovery of gender classification rules for social network data using data mining algorithms. In: Proceedings of the IEEE International Conference on Computational Intelligence and Computing Research (ICCIC 2011); Kanyakumari, India (2011)

19. Yuvaraj, N., Chang, V., Pinagapani, A., Kannan, S., Dhiman, G., Rajan, A.R.: Automatic detection of cyberbullying using multi-feature based artificial intelligence with deep decision tree classification, Elsevier. Comput. Electric. Eng. **92** (2021)

20. Kumar, B.M., Perumal, R.S., Nadesh, R.K., Arivuselvan, K.: Type 2: diabetes mellitus prediction using Deep Neural Networks classifier. Int. J. Cogn. Comput. Eng. **1**, 55–61 (2020)

21. Strzelecka, A., Zawadzka, D.: Application of classification and regression tree (CRT) analysis to identify the agricultural households at risk of financial exclusion. Procedia Comput. Sci. **192**, 4532–4541 (2021)

22. Sharma, S., Agrawal, J., Sharma, S.: Classification through Machine Learning Technique: C4.5 Algorithm based on Various Entropies No 16 (2013)

23. Domingos, P.: MetaCost: a general method for making classifiers cost-sensitive. In: Proceedings of the Fifth International Conference on Knowledge Discovery and Data Mining, pp. 155–164. ACM Press, San Diego, CA (1999)
24. Chawla, N.V., Japkowicz, N., Kolcz, A. (eds.) Special Issue on Learning from Imbalanced Datasets. SIGKDD, vol. 6, issue 1. ACM Press (2004)
25. Zubek, V.B., Dietterich, T.: Pruning improves heuristic search for cost-sensitive learning. In: Proceedings of the Nineteenth International Conference of Machine Learning, pp. 27–35, Morgan Kaufmann, Sydney, Australia (2002)
26. Madadipouya, K.: A new decision tree method for data mining in medicine. Adv. Comput. Intell. Int. J. 2(3) (2015)

How far can Deep Learning Improve Arabic Part of Speech Tagging?

Mohamed Bouzahir(✉) 🆔, Abdelkaher Ait Abdelouahad🆔, and Mohamed Nabil🆔

LAROSERI Laboratory, Computer Science Department, Faculty of Sciences,
University of Chouaib Doukkali, El Jadida, Morocco
mbouzahir@gmail.com

Abstract. The use of Deep Learning (DL) in Natural Language Processing (NLP) has seen a significant growth in the last few years. Part of Speech (POS) tagging is an important element for many NLP applications, including machine translation, sentiment analysis, and text summarization. It consists in identifying the very likely tag (noun, verb or particle, adverbs, adjectives etc.) for each word in a given text. The goal of this research is to conduct a systematic literature review of deep learning methods applied to Arabic POS tagging for the last two decades. The review was conducted using the Preferred Reporting Items for Systematic Reviews and Meta-Analyses (PRISMA) framework. More than 4000 papers were reviewed to extract all DL approaches used to develop POS taggers for the Arabic language. After multiple exclusion steps 12 articles were selected for a full review. Results show that Long Short-Term Memory (LSTM) and its extension Bidirectional LSTM (Bi-LSTM) models are the most used DL techniques for Arabic POS tagging, and they give better results according to the reviewed papers.

Keywords: Neural network · Deep learning · Arabic POS-tagger · ANLP · PRISMA

1 Introduction

Part of speech tagging is one of the critical steps in natural language processing. It is the process of finding the very likely grammatical tag (nouns, adverbs, verbs, adjectives. etc.) for a given word, this process is also known as grammatical tagging. In Arabic, this task encounters many problems because the Arabic language has a very rich concord system [1]. In addition, agglutination and the lack of diacritics in Arabic writing make the task even harder [1].

Like other natural language processing tasks, we distinguish three categories of part-of-speech (POS) tagging methods:

Rule-based method [2]. It consists of specifying the language rules to the system that will apply these rules to predict the POS tag. For example, we can tell the system that a word ending with "ة"(Taa marbouta) is a noun. The system can then identify nouns based on this rule. However, However, given the complexities of human language, it is almost impossible to avoid ambiguity when using just language rules [1].

© Springer Nature Switzerland AG 2022
M. Fakir et al. (Eds.): CBI 2022, LNBIP 449, pp. 206–215, 2022.
https://doi.org/10.1007/978-3-031-06458-6_17

Stochastic/probabilistic methods [3]. Instead of giving the rules, only an annotated corpus is required to train the system, then using some statistical algorithms the system can predict the grammatical tag of the word. Because of the randomness of statistics, some researchers tried hybrid methods, combining rule-based and statistical methods to gain the benefits of the two methods [4].

In the last two decades researchers started utilizing Deep Learning (DL) approaches for many NLP tasks including POS tagging. In fact, the availability of a huge amount of data and performance increase of computer hardware, made computer training, using artificial neural networks, more affordable. Hence, the computer generates its rules and predict the right tag for the word.

In this paper, we conduct a literature review of deep learning methods applied to POS tagging for the Arabic language. The review is conducted using the PRISMA framework, mostly used in medical researches. The remaining of this paper is divided into two sections. The first section discusses non-DL approaches applied to POS tagging for the Arabic language, while the second section contains a review of DL methods used for POS tagging of Arabic. Finally, a conclusion ends the paper.

2 Non-DL Approaches

Before the last two decades, deep learning methods were rarely used, and the approach has not provided good results compared to stochastics and hybrid approaches [5]. The first POS tagger for Arabic was made for commercial purposes by companies like Xerox. Xerox developed their system in 2001 using grammar rules in a Finite-State machine [6]. However, Parallel research was conducted. In 2002, Maamouri and Cieri [7] proposed a rule-based tagger based on the output of the morphological analyzer of Tim Buck-walter [8]. Based on statistics and rule-based techniques, Shereen Khoja [9] developed Automatic Arabic Tagger (APT). This was the first tagger to use a real Arabic grammar tagset. Using a manually constructed corpus of 3000 tagged words, Freeman [10] in 2001 used Machine learning and developed Brill's POS-tagger for the Arabic language using a corpus of 146 tags. Diab et al. [11], developed in 2004 a POS-tagger for the Arabic language based on Support Vector Machine (SVM). They reach an accuracy of 95.49% on the Arabic Tree Bank (ATB) corpus. Two years later, Guiassa [12] developed, a new POS tagger for the Arabic language based on a hybrid method, combining a rule-based method and memory-based learning. The accuracy was 86%. Another part-of-speech tagger was developed in 2006 by Shamsi and Guessoum [13] using Hidden Markov Model (HMM). They claim an accuracy of 97% on the ATB corpus. Alqrainy [14], developed Arabic Morphosyntactic Tagger (AMT) using a rule-based technique based on Arabic pattern, lexical and contextual rules. In 2009, Elhadj [15] proposed a hybrid POS-tagger combining morphological analysis output with HMM technique using a corpus of 56312 words.

Mohamed and Kübler [16], in 2010, proposed two approaches for Arabic POS-tagger. The first one is without segmentation of the words and the second method is with word segmentation using machine learning segmenter. An accuracy of 94.74% was reached using the ATB dataset. Three years later Ali and Jarray [17] developed a new approach for Arabic POS tagging using a Genetic algorithm. Hadni et al. [4] in 2013 developed a new POS Tagger for Arabic taking into consideration Non-Vocalized Text using HMM combined with a rule-based method.

Kadim and Lazrek [18] suggested a stochastic method based on Bidirectional HMM for Arabic POS tagging. Ababou and Mazroui [5] reached 94.02% accuracy on Nemlar corpus using a hybrid approach combining HMM and AlKhalil morphology analyzer output [19]. Hjouj Btoush et al. [20] proposed a rule-based approach composed of two phases, a lexicon one followed by a morphological one. Abdulkareem and Tiun [21] compared three stochastic methods for POS tagging of Arabic dialects tweets: key nearest neighbour, naïve Bayes and decision tree learning Iterative Dichotomiser 3 (ID3). They concluded that the Decision tree ID3 approach is the best suited for Arabic dialects.

Ba-Alwi et al. [22] compared three stochastic methods: HMM with the prefix guessing, HMM with linear interpolation guessing and Trigrams'n'Tags (TnT), using morpheme and word. HMM with linear interpolation guessing method shows better results over TnT.

Kadim and Lazrek [23] proposed in 2018 another approach combining two HMMs working in parallel and two corpora. In the same year, Aliwy and Raza [24] reached 97.7% accuracy on the KALIMAT corpus combining n-gram and HMM. Labidi [25] in 2019 proposed a new approach combining two taggers with two different methods, one using maximum entropy and another with a statistical/rule-based method.

Before 2006 we noticed a lack of use of deep learning to solve POS tagging task in the Arabic language, the accuracy is between 88% and 97% for most Non-DL-based taggers. Different Datasets such as ATB, NEMLAR and Quranic Arabic Corpus were used. In the next section, we will conduct our review about the use of deep learning techniques for Arabic POS tagging following the PRISMA framework. Table 1 resumes the list of cited works with year, methodology and accuracy. We noticed that for some approaches the accuracy value was not mentioned.

Table 1. Non-DL methods applied to Arabic POS tagging.

Author	Year	Dl technique	Accuracy
	2001	Hybrid	86%
Freeman [10]	2001	Machine learning	–
Maamouri and Cieri [7]	2002	Rule-based	96%
Mona Diab [11]	2004	SVM	–
Tlili-Guiassa [12]	2006	Hybrid	86%
Shamsi and Guessoum [13]	2006	Statistical	97%
Shihadeh Alqrainy [14]	2008	Rule-based	91%
Elhadj [15]	2009	HMM	96%
Mohamed and Kübler [16]	2010	Segmentation-based tagging	93.47
Mohamed and Kübler [16]	2010	Whole word-based tagging	94.74%

(*continued*)

Table 1. (*continued*)

Author	Year	Dl technique	Accuracy
Ali and Jarray [17]	2013	Genatic Algorithm	–
Kadim and Lazrek [18]	2016	Bidirectional HMM	–
Ababou and Mazroui [5]	2016	Hybrid (AlKhalil + HMM)	–
Hjouj Btoush et al. [20]	2016	Rule-based	–
Abdulkareem and Tiun [21]	2017	Decision tree ID3	–
Ba-Alwi et al. [22]	2017	HMM	–
Kadim and Lazrek [23]	2018	HMM	–
Aliwy and Raza [24]	2018	N-gram + HMM	–
Labidi [25]	2019	Entropy + statistical/rules-based	85%–90%

3 Deep Learning for POS Tagging

Deep learning refers to the use of ANN (Artificial Neural Network) in the machine learning systems. In 1943 McCulloch & Pitts invented the first mathematical model of neuron network that imitate the biological neuron [26]. Fifteen years later in 1958 Rosenblatt proposed a developed model named perceptron [27]. Because of the weak material and algorithm at that time its usage was very limited. However, in the recent two decades, the use of ANN increased. Now ANN is used to solve a lot of problems such as clustering [28], classification [29], regression [30], machine translation [31] and more.

Many DL techniques have been applied to POS tagging for Indo-European languages. In 1994, Schmid got 96.22% accuracy for his POS-tagger using MLP [32]. In 2015 Wang et al. [33] reached 97.26% accuracy using BiLSTM. Gated Recurrent Unit (GRU) and other types of recurrent neural network (RNN) have been used to solve the POS tagging task. Inspired by results in Indo-European languages researchers in Arabic natural language processing started using DL techniques to solve many tasks including POS tagging. In the next subsection, we describe the method we used for our review.

3.1 Method

This review was conducted following the PRISMA method [34]. PRISMA instructions help researchers improve the reporting of systematic reviews by following a checklist of 27 items. The items are related to the content of the review starting from the title and abstract of the paper until the conclusion. In this process, we have to mention the number of papers that were excluded or included in the review and the criteria of exclusion.

Figure 1 illustrates the flow diagram of the PRISMA and its application in our review. The process consists of three steps: identification, screening and inclusion. The primary requirement for inclusion in this review was that the article has to introduce a POS tagging system for the Arabic language using a deep learning method. The article is excluded if it is not in English or full text is not available. Based on a python crawler script originally created by Wally Liu [35]. We developed our script [36] to collect articles from selected scientific engines. We used the top known academic search engines like Microsoft Academic, Google Scholar, BASE and CORE to retrieve all documents all papers related to Arabic POS tagging using deep learning methods. We defined the search terms (Arabic POS deep learning, Arabic POS Neural Network, POS tagging, POS tagger, Arabic deep learning). The searches were limited to publication dates between 2006 and June 2021. Results give 3932 papers from Microsoft Academic, 2299 from Google scholar, 845 from Semantic scholar, 73 from Core. The results are then imported into a MySQL database to exclude duplicated and not eligible papers. Using SQL queries, we removed 3462 duplicated papers and 3562 ineligible papers not having **POS/Part-of-/part of speech** and **deep learning/neural network** in the title and abstract.

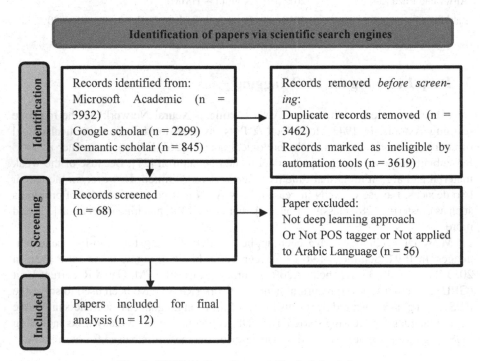

Fig. 1. PRISMA flow diagram of included articles.

In the screening step, we manually reviewed the 68 papers resulting from the first step. We eliminated 57 papers that did not provide a deep learning approach for Arabic POS tagging. At the end of the screening 12 articles were included in the final review using full-text versions.

3.2 Discussion

In this subsection, we will discuss the results of the 12 papers included in the final step of our review. Table 2 illustrates the summary statistics of the selected papers including the used DL technique, the calculated accuracy/ F1 score, the dataset name, when given, or the number of words of training and testing and the size of the tagset.

The 12 papers chosen for full-text review were focusing on developing a part of speech system for the Arabic language using a deep learning approach and techniques such as Bi-LSTM-CRF (n = 4), Bi-LSTM (n = 4), LSTM (n = 1), BPNN (n = 2), MLP (n = 1).

Different tagsets and corpora were used in the training and testing. The approaches applied to Arabic POS tagging gave different results even with the same techniques [47, 48].

This paper presents a systematic review of POS tagging techniques using deep learning methods, 12 papers were reviewed from more than 68 screened titles and abstracts. More than 70% of the papers are using a modified Bi-LSTM model either by adding some processing layers or a Conditional random field (CRF) layer.

Table 2. Comparison of articles methods and accuracy.

Author	Year	Dl technique	Accuracy	Dataset	Size of tagset
Yousif [37]	2006	Multilayer perceptron (MLP)	99.99%	50,000 words	131 tags
Muaidi [38]	2014	Backpropagation Neural Network (BPNN)	90.21%	24,810 words	161 tags
Plank et al. [39]	2016	Bi-LSTM	97.22%	Arabic Treebank (ATB) [40]	70 tags
Darwish et al. [41]	2017	Bi-LSTM	96.2%	WikiNews 18,300 words	18 tags
Abumalloh et al. [42]	2018	BPNN	89.04%	20,620 words	18 tags
Alharbi et al. [43]	2018	Bi-LSTM	91.2%	6,844 tokens	21 tags
AlGhamdi and Diab [44]	2019	Bi-LSTM-CRF	92.90%	ATB 174,967 words	70 tags
Alrajhi and ELAffendi [45]	2019	LSTM at Word level	99.18%	124,002 words	87 tags
		LSTM at morpheme level	99.72%	77,190 words	34 tags

(continued)

Table 2. (*continued*)

Author	Year	Dl technique	Accuracy	Dataset	Size of tagset
Attia et al. [46]	2019	Bi-LSTM-CRF	92.38%	–	28 tags
Darwish et al. [47]	2020	Bi-LSTM-CRF	92.4%	35951 words	24 tags
Younes and Weeds [48]	2020	Bi-LSTM-CRF	95.4%–98.00%	WikiNews, PADT, Al Mushaf	
AlKhwiter and Al-Twairesh [49]	2021	Bi-LSTM	96.5%	75,677 words	22 tags

4   Remarks and Challenges

In this subsection, we examine the researcher's statements on POS tagging for the Arabic language based on the methodology and performance standards they developed. It also indicates future trends as well as forthcoming gaps and problems for researchers to develop an effective and efficient Arabic POS tagger.

4.1   Remarks

In this literature review, we witnessed the use of different small corpora and low numbers of tags, which make the comparison of applied method inaccurate and claimed accuracies may significantly decrease for other large corpora. Some authors attempted to compare their DL approach with other approaches such as SVM [49]. In most circumstances, the DL model outperformed the other method. The performance of deep learning models applied to POS tagging can be improved by adding some linguistic features [46, 49]. The common challenge is the absence of Arabic annotated corpus and datasets dedicated to POS tagging.

4.2   Challenges

Absence of Standard Dataset

The use of an adequate corpus, specifically in the training phase determine the level of success of any DL based POS tagging. For traditional machine learning algorithms, good results can be achieved relying just on a limited dataset but deep learning approaches need a large corpus to outperform other machine learning techniques. In this study, we noticed a lack of a standard corpus for POS taggers for the Arabic language, which make any comparison of the given approaches inaccurate.

To improve the efficiency of Arabic POS tagger, the creation of an adequate and standard corpus is of a high priority for researchers. Such a tedious task will require concerted efforts and more cooperation between linguists and computer scientists. To reduce Arabic resource scarcity, such corpus should be freely available to the research community.

Resource Requirement
Deep learning algorithms applied to POS tagging are time-consuming and need High-Performance Computing Resources due to the large size of the training corpus, which makes the learning phase very challenging, especially for the Arab countries where such computer resources are indeed very costly.

5 Conclusion

In this paper, we tried to answer the question: how far DL can improve Arabic POS tagging? First, we have reported the most relevant works related to Arabic POS tagging, using only classical methods: rule-based, statistics/stochastics or hybrid ones. Then, we have conducted a systematic review of deep learning-based Arabic POS Tagging, from 2006 to June 2021, using the PRISMA framework. This framework offers a methodical approach to make an overview of all the papers related to our subject. After a series of automatic and manual steps, only papers proposing a deep learning approach to POS tagging of the Arabic language were selected. Results of POS Arabic Tagger using DL are significantly higher than those using only classical methods. Also, we noticed extensive use of bi-LSTM and CRF but different accuracies which might be because of the size and quality of the different used datasets. This does not mean that one has to abandon classical methods in favour of DL based ones. Instead, why not make use of both methods?

As a future work, we intend to apply the cited DL methods on the same corpus and compare the results using the same measuring technique to get a better overview of the effectiveness of each approach.

References

1. Farghaly, A., Shaalan, K.: Arabic natural language processing: challenges and solutions. ACM Trans. Asian Lang. Inf. Process. **8**(4), 14:1–14:22 (2009)
2. Brill, E.: A simple rule-based part of speech tagger. In: Proceedings of the Third Conference on Applied Natural Language Processing, pp. 152–155. USA (1992)
3. Martinez, A.R.: Part-of-speech tagging. WIREs Comput. Stat. **4**(1), 107–113 (2012)
4. Hadni, M., Ouatik El Alaoui, S., Lachkar, A., Meknassi, M.: Hybrid part-of-speech tagger for non-vocalized Arabic text. Int. J. Nat. Lang. Comput. **2**, 1–15 (2013)
5. Ababou, N., Mazroui, A.: A hybrid Arabic POS tagging for simple and compound morphosyntactic tags. Int. J. Speech Technol. **19**(2), 289–302 (2015). https://doi.org/10.1007/s10 772-015-9302-8
6. Beesley, K.: Finite-State Morphological Analysis and Generation of Arabic at Xerox Research: Status and Plans in 2001 (2001)
7. Maamouri, M., Cieri, C.: Resources for Arabic Natural Language Processing, p. 14 (2002)
8. Buckwalter, T.: Buckwalter Arabic morphological analyzer version 1.0, Linguist. Data Consort. University PA (2002)

9. Khoja, S.: APT: Arabic part-of-speech tagger. In: Proceedings Student Workshop NAACL, pp. 20–25 (2001)
10. Freeman, A.: Brill's POS tagger and a Morphology parser for Arabic (2001)
11. Diab, M., Hacioglu, K., Jurafsky, D.: Automatic tagging of Arabic text: from raw text to base phrase chunks. In: Proceedings of HLT-NAACL (2004)
12. Tlili-Guiassa, Y.: Hybrid method for tagging Arabic text. J. Comput. Sci. **2**, 245–248 (2006)
13. Shamsi, F., Guessoum, A.: A hidden Markov model-based POS tagger for Arabic (2006)
14. Alqrainy, S., Muaidi, H., Ayesh, A.: Pattern-based algorithm for Part-of-Speech tagging Arabic text, p. 124 (2008)
15. Elhadj, Y.: Statistical part-of-speech tagger for traditional Arabic texts. J. Comput. Sci. **5**, 794–800 (2009)
16. Mohamed, E., Kübler, S.: Is Arabic Part of Speech Tagging Feasible Without Word Segmentation? pp. 705–708 (2010)
17. Ali, B.B., Jarray, F.: Genetic approach for Arabic part of speech tagging. arXiv:13073489 Cs (2013)
18. Kadim, A., Lazrek, A.: Bidirectional HMM-based Arabic POS tagging. Int. J. Speech Technol. **19**(2), 303–312 (2015). https://doi.org/10.1007/s10772-015-9303-7
19. Boudlal, A., Lakhouaja, A.: Alkhalil morpho sys1: a morphosyntactic analysis system for Arabic texts. In: International Arab Conference on Information Technology, pp. 1–6 (2010)
20. Hjouj Btoush, M., Alarabeyyat, A., Olab, I.: Rule based approach for Arabic part of speech tagging and name entity recognition. Int. J. Adv. Comput. Sci. Appl. **7** (2016)
21. Abdulkareem, M., Tiun, S.: Comparative analysis of ML POS on Arabic tweets, vol. 95, pp. 403–411 (2017)
22. Ba-Alwi, F., Albared, M., Al-Moslmi, T.: Choosing the optimal segmentation level for POS tagging of the Quranic Arabic. Br. J. Appl. Sci. Technol. **19**, 1–10 (2017)
23. Kadim, A., Lazrek, A.: Parallel HMM-Based Approach for Arabic Part of Speech Tagging, vol. 15, no. 2, p. 11 (2018)
24. Aliwy, A.H., Raza, D.A.A.: Part of speech tagging for Arabic long sentence. Int. J. Eng. Technol. **7**(3.27)+(Art. no. 3.27) (2018)
25. Labidi, M.: New combined method to improve Arabic POS tagging. J. Auton. Intell. **1**, 23 (2019)
26. McCulloch, W.S., Pitts, W.: A logical calculus of the ideas immanent in nervous activity. Bull. Math. Biophys. **5**(4), 115–133 (1943)
27. Rosenblatt, F.: The perceptron: a probabilistic model for information storage and organization in the brain. Psychol. Rev. **65**(6), 386–408 (1958)
28. Aljalbout, E., Golkov, V., Siddiqui, Y., Strobel, M., Cremers, D.: Clustering with Deep Learning: Taxonomy and New Methods. arXiv:1801.07648 Cs Stat (2018)
29. Tabar, Y.R., Halici, U.: A novel deep learning approach for classification of EEG motor imagery signals. J. Neural Eng. **14**(1), 016003 (2016)
30. Qiu, X., Zhang, L., Ren, Y., Suganthan, P.N., Amaratunga, G.: Ensemble deep learning for regression and time series forecasting. In: 2014 IEEE Symposium on Computational Intelligence in Ensemble Learning (CIEL), pp. 1–6 (2014)
31. Singh, S.P., Kumar, A., Darbari, H., Singh, L., Rastogi, A., Jain, S.: Machine translation using deep learning: an overview. In: 2017 International Conference on Computer, Communications and Electronics (Comptelix), pp. 162–167 (2017)
32. Schmid, H.: Part-of-Speech Tagging with Neural Networks. arXiv:cmp-lg/9410018 (1994)
33. Wang, P., Qian, Y., Soong, F.K., He, L., Zhao, H.: A Unified Tagging Solution: Bidirectional LSTM Recurrent Neural Network with Word Embedding. arXiv:1511.00215 Cs (2015)
34. Moher, D., Liberati, A., Tetzlaff, J., Altman, D.G., Group, T.P.: Preferred reporting items for systematic reviews and meta-analyses: the PRISMA statement. PLOS Med. **6**(7), e1000097 (2009)

35. Liu, W.: wallyliu/PACInterview (2017)
36. Bouzahir, M.: bounmed/pythonCrawlers (2021). https://github.com/bounmed/pythonCrawlers. Accessed 17 July 2021
37. Yousif, J.: Design and implement an automatic neural tagger based Arabic language for NLP applications. Asian J. Inf. Technol. **5**, 784–789 (2006)
38. Muaidi, H.: Levenberg-marquardt learning neural network for part-of-speech tagging of Arabic sentences. WSEAS Trans. Comput. **13**, 300–309 (2014)
39. Plank, B., Søgaard, A., Goldberg, Y.: Multilingual Part-of-Speech Tagging with Bidirectional Long Short-Term Memory Models and Auxiliary Loss. arXiv:1604.05529 Cs (2016)
40. Bies, A., Maamouri, M.: Penn Arabic Treebank guidelines (2009)
41. Darwish, K., Mubarak, H., Abdelali, A., Eldesouki, M.: Arabic POS tagging: don't abandon feature engineering just yet. In: Proceedings of the Third Arabic Natural Language Processing Workshop, pp. 130–137. Valencia, Spain (2017)
42. Abumalloh, R., Muaidi, H., Ibrahim, O., Abu-Ulbeh, W.: Arabic part-of-speech tagger, an approach based on neural network modelling. Int. J. Eng. Technol. **7**, 742 (2018)
43. Alharbi, R., Magdy, W., Darwish, K., AbdelAli, A., Mubarak, H.: Part-of-Speech Tagging for Arabic Gulf Dialect Using Bi-LSTM. Presented at the LREC 2018, Miyazaki, Japan (2018)
44. AlGhamdi, F., Diab, M.: Leveraging Pretrained Word Embeddings for Part-of-Speech Tagging of Code Switching Data (2019)
45. Alrajhi, K., ELAffendi, M.A.: Automatic Arabic Part-of-Speech Tagging: Deep Learning Neural LSTM Versus Word2Vec (2019)
46. Attia, M.A., Samih, Y., El-Kahky, A., Mubarak, H., Abdelali, A., Darwish, K.: POS Tagging for Improving Code-Switching Identification in Arabic (2019)
47. Darwish, K., et al.: Effective multi-dialectal Arabic POS tagging. Nat. Lang. Eng. **26**(6), 677–690 (2020)
48. Younes, A., Weeds, J.: Embed More Ignore Less (EMIL): enriched representations for Arabic NLP (2020)
49. AlKhwiter, W., Al-Twairesh, N.: Part-of-speech tagging for Arabic tweets using CRF and Bi-LSTM. Comput. Speech Lang. **65**, 101138 (2021)

Optimization and Dynamic Programming

Analysis of Several Algorithms for DOA Estimation in Two Different Communication Models by a Comparative Study

Hassan Ougraz[1]([✉]) [iD], Said Safi[1] [iD], and Miloud Frikel[2] [iD]

[1] LIMATI Laboratory, Department of Mathematics and Computer Sciences, Sultan Moulay Slimane University, Beni Mellal, Morocco
{hassan.ougrazfpb,said.safi}@usms.ac.ma
[2] Automatic Laboratory, UNICAEN, ENSICAEN, Normandie University, Caen, France
miloud.frikel@ensicaen.fr

Abstract. Direction-of-arrival (DOA) is a critical parameter in array signal processing. The classical DOA estimation approaches are inefficient in applications such as underwater array processing because it consists of a high number of snapshots. In the recent advancement of technology, wideband signals are preferable to narrowband signals. Wideband signals can estimate DOAs more effectively with fewer side lobes and antenna elements. In this study, we compare the performance of DOA estimation for narrowband and broadband signals by evaluating the angular spectrum of the multiple signal classification (MUSIC) algorithm and the Capon method. We will estimate the spectral position using various parameters such as the number of antennas, and signal-to-noise ratio (SNR). In addition, we will look for the spectral peak location and calculate the final DOA. As a result, studying and analyzing wideband signals is crucial, particularly in applications such as 5G MIMO systems.

Keywords: DOA estimation · Narrowband signals · Wideband signals · Uniform linear array · MUSIC algorithm · Capon algorithm

1 Introduction

Direction-of-arrival (DOA) is a significant issue in a variety of domains, including sonar, radar, acoustic tracking, array signal processing, wireless communication, and many other applications [1–7]. DOA estimation methods use an antenna array to determine the direction of a received signal. Numerous techniques are provided under certain assumptions, such as a large number of snapshots, the signal-to-noise ratio (SNR) is high and the signal sources are uncorrelated. Multiple signal classification (MUSIC) and Capon algorithms are two of the most famous approaches have been developed in the last four decades [8–11].

The fundamental concept behind the MUSIC technique [8] is to construct a correlation matrix from the collected signal of a uniform linear array and

© Springer Nature Switzerland AG 2022
M. Fakir et al. (Eds.): CBI 2022, LNBIP 449, pp. 219–230, 2022.
https://doi.org/10.1007/978-3-031-06458-6_18

decompose it using eigenvalue decomposition to obtain signal and noise sub-spaces; two perpendicular components. By decreasing the output power, the Capon minimum variance approach [9] enhances the performance of the conventional beamforming method and reduces the influence of noise [12]. These approaches are not directly applicable to wideband signals, but the performance of MUSIC in terms of numerical convergence is fast, and there is no requirement for a prior guess of DOA.

Much research has been done in narrowband signals for DOA estimation over the last 20–30 years, which it requires a large number of snapshots and provides super-resolution DOA estimation for narrowband uncorrelated signals [13]. Furthermore, a wideband signal is frequently employed in information systems, particularly in 5G MIMO systems [14,15]. As a result, among the several directions of arrival algorithms proposed before, MUSIC and Capon are commonly employed since they give a super-resolution and greater accuracies [16].

In this paper, we present a comparison of DOAs estimation for narrowband and wideband scenarios, using the MUSIC and Capon algorithms in the one-dimensional (1D) stationary condition.

The remainder of the paper is as follows. In Sect. 2 the problem formulation and the fundamentals of DOA estimation will be presented, followed by Sect. 3 that explains the narrowband and wideband methods. Section 4 shows the performance evaluation and discussion for each algorithm. Finally, Sect. 5 involves the conclusion and perspective.

2   Problem Formulation

As shown below in Fig. 1 we have a uniform linear array (ULA) with N omni-directional sensors that are equally spaced apart by a distance d.

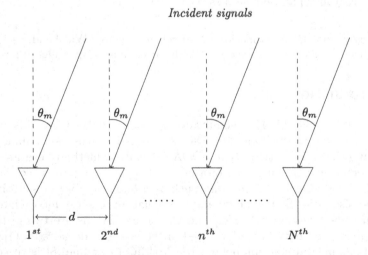

Incident signals

Fig. 1. Geometry of an N element linear array

2.1 Narrowband Signal Model

Let us consider an antenna array with N elements that receives signals from M distant field point sources ($M < N$). The goal is then to separate the M known sources by using the signals received by the array antennas and the information provided at reception. On the array, the waves released by the sources arrive from different directions ($\theta_1, \theta_2, \ldots, \theta_M$). Following that, the received signals are noisy linear combinations of the source signals. By making K observations, the receiving vector at the output of the antenna array corresponding to an observation is expressed according to the following model:

$$x_m(t) = \sum_{m=1}^{M} a(\theta_m)s_m(t) + n(t) = A(\theta)s(t) + n(t) \tag{1}$$

where $n(t)$ is a white noise array and zero mean complex Gaussian, $A(\theta)$ is a matrix consisting of M directing arrays, and $a(\theta_m)$ is the directing vector corresponding to the DOA of the mth signal source, defined as:

$$a(\theta_m) = \left[1,\ e^{-jv_m},\ \ldots,\ e^{-j(N-1)v_m}\right]^T \tag{2}$$

where $[.]^T$ denotes the transposed operator, v_m is the phase shift between the antenna array items and is represented as:

$$v_m = 2\pi \left(\frac{d}{\lambda}\right) \sin \theta_m \tag{3}$$

where λ represents the wavelength.

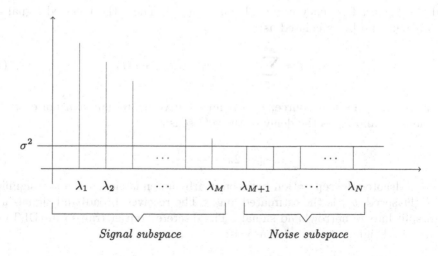

Fig. 2. Representation of eigenvalues in descending order

The correlation matrix, in narrowband scenario, is expressed by:

$$
\begin{aligned}
R_{xx}(t) &= E\big[x(t)x^H(t)\big] \\
&= A(\theta)R_{ss}(t)A^H(\theta) + \sigma^2 I,
\end{aligned}
\tag{4}
$$

where $E[.]$ represents the expectation value operator, $(.)^H$ is the complex conjugate transpose, $R_{ss}(t)$ is the covariance matrix of the signal vector $(M \times M)$, σ^2 is the same noise power for each sensor, and I is the identity matrix $(N \times N)$.

Since the $R_{xx}(t)$ matrix is hermitian and positive, its values are real and positive. Its non-zero eigenvalues are classified in the following order:

$$
\mu_1 \geq \mu_2 \geq \ldots \geq \mu_M
\tag{5}
$$

The N eigenvalues of the $R_{xx}(t)$ matrix can be written as:

$$
\begin{aligned}
\lambda_m &= \mu_m + \sigma^2, && 1 \leq m \leq M \\
\lambda_m &= \sigma^2, && M+1 \leq m \leq N
\end{aligned}
\tag{6}
$$

where λ_m are the N eigenvectors of $R_{xx}(t)$ associated to the N eigenvalues.

2.2   Wideband Signal Model

Let us assume that the M wideband sources $(M < N)$ are identifiable or can be calculated [17,18], with identical bandwidth that is impinging on the number of array sensors from directions $(\theta_1, \theta_2, \ldots, \theta_M)$. The Fig. 1 shows the ULA configuration for DOA estimation.

The objective is to estimate the DOA of M wideband source signals using a ULA of N antennas. It is also considered that all signals are uncorrelated and exist within the bandwidth between the lowest frequency of signal sources (f_L) and the highest frequency of signal sources (f_H). Then, the received signal at n^{th} element can be calculated as:

$$
x_m(t) = \sum_{m=1}^{M} s_m\left(t - v_n \sin\theta_m\right) + n(t),
\tag{7}
$$

where $s_m(t)$ is the mth source, $n(t)$ is an additive white gaussian noise at the nth element, and v_m is the delay of the mth sensor,

$$
v_n = 2\pi\frac{(n-1)d}{c},
\tag{8}
$$

where d denotes the separation between nearby antenna elements and c signifies the light speed. θ_m is the estimated angle. The received broadband signals are then split into K narrowband signals. The discrete Fourier transform (DFT) of the received signal at the nth sensor is:

$$
x_m(f) = \sum_{m=1}^{M} s_m(f)e^{-jfv_n \sin\theta_m} + n(f)
\tag{9}
$$

Then, the output signals, in Fourier domain, can be expressed in vector form as follows:

$$x(f_k) = A(f_k, \theta)s(f_k) + n(f_k), \quad 1 \leq k \leq K, \tag{10}$$

where $f_k \in]f_L, f_H[$, for $k = 1, 2, \ldots, K$.

$$A(f_k, \theta) = [a(f_k, \theta_1), \ a(f_k, \theta_2), \ \ldots, \ a(f_k, \theta_M)], \tag{11}$$

$$a(f_k, \theta_m) = \left[1, \ e^{-jf_k v_1 \sin \theta_m)}, \ \ldots, \ e^{-jf_k v_{N-1} \sin \theta_m}\right]^T \tag{12}$$

The correlation matrix $R_{xx}(f_k)$, in wideband scenario, is calculated by the following equation:

$$\begin{aligned} R_{xx}(f_k) &= E[x(f_k)x^H(f_k)] \\ &= A(f_k, \theta)R_{ss}(f_k)A^H(f_k, \theta) + \sigma_n^2 I, \end{aligned} \tag{13}$$

where $R_{ss}(f_k) = E[s(f_k)s^H(f_k)]$. Assuming that the M signal sources are uncorrelated, $R_{ss}(f_k)$ has full order, then the matrix of signal subspaces $U_s(f_k)$ and teh array of noise subspaces $U_n(f_k)$ at frequency f_k may be generated using the covariance matrix's eigenvalue decomposition (EVD) as:

$$U_s(f_k) = [\lambda_{k,1}, \lambda_{k,2}, \ldots, \lambda_{k,M}], \tag{14}$$

$$U_n(f_k) = [\lambda_{k,M+1}, \lambda_{k,M+2}, \ldots, \lambda_{k,N}], \tag{15}$$

where $(\lambda_{k,1}, \ldots, \lambda_{k,N})$ are the perpendicular eigenvectors of $R_{xx}(f_k)$, sorted in decreasing order by their eigenvalues.

3 DOA Estimation Algorithms

In this section, the MUSIC method proposed by Schmidt [8] and the Capon method proposed by Capon [9] are evaluated to detect and estimate the direction of the incident signals.

3.1 Narrowband MUSIC

Schmidt [8] independently has proposed a simple and well-known approach known as multiple signal classification (MUSIC) that helps to calculate the angles of arrival using a pseudo-spectrum determined from the signal and noise subspaces of the observations. However, when two directions have extremely close angles, MUSIC spatial does not always enable them to be separated. In fact, the spatial resolution is proportional to the number of antennas N. The number of directions that can be localized is also affected by this last parameter. As a result, the resolution of M directions requires a number of antennas bigger than or equal to $(M + 1)$. To determine the different directions of arrival by applying the MUSIC method, follow these steps:

- Search for eigenvectors and eigenvalues of the correlation matrix.
- Classification of these eigenvalues according to the descending values and identification of the eigenvectors associated with the most important eigenvalues; these vectors form a base of the signal subspace. The remaining eigenvectors constitute a base of the noise subspace.
- Construction of a family of vectors parameterized by the angle of arrival having a theoretical relative phase state imposed by the geometry of the array.
- Projection of this family of vectors on the noise subspace, and the values of the directions of arrival for which a relative minimum is required are determined.

The starting hypothesis is that the covariance matrix $R_{xx}(t)$, Eq. (4), is not singular, this is interpreted physically by the fact that all the sources are decorrelated.

The angular spectral function computed using the narrowband MUSIC approach is used to find the values of θ for which this function is maximal and is defined as:

$$P(\theta) = \frac{1}{a^H(\theta)U_n(t)U_n^H(t)a(\theta)} \tag{16}$$

The MUSIC pseudo-spectrum provides peaks matching to the exact directions of arrival of the waves but does not provide information on the power of the sources.

3.2 Narrowband Capon

The Capon technique is also known as MVDR (minimum variance distortionless response) [9]. The idea is to reduce the system's average power with the exception of one that pointing in the wanted signal location:

$$\min_{\omega} \omega^H R_{xx}(t)\omega \quad \text{subject to} \quad |\,\omega^H a(\theta)\,| = 1 \tag{17}$$

The weighted array $\omega(\theta)$ would be [9]:

$$\omega(\theta) = \frac{R_{xx}^{-1}(t)a(\theta)}{a^H(\theta)R_{xx}^{-1}(t)a(\theta)} \tag{18}$$

The narrowband Capon technique will investigate all degrees for the direction with the highest measured power gained:

$$Q(\theta) = \omega^H(\theta)R_{xx}(t)\omega(\theta) = \frac{1}{a^H(\theta)R_{xx}^{-1}(t)a(\theta)} \tag{19}$$

3.3 Wideband MUSIC

This method is known as wideband MUSIC because the same narrowband MUSIC approach is used for all frequency bins at the same time. Wideband MUSIC uses narrowband signal subspaces approches to every frequency range separately [19].

The wideband MUSIC computes the DOA of wideband sources by employing the formula:

$$P(\theta) = \frac{1}{\sum_{k=1}^{K} a^H(f_k,\theta)U_n(f_k)U_n^H(f_k)a(f_k,\theta)} \tag{20}$$

The noise subspace matrix $U_n(f_k)$ is estimated from the spatial correlation matrix $R_{xx}(f_k)$ in Eq. (15). While the DOAs predicted by the Eq. (20) are average values of every frequency range's output, the wrong estimations from a single frequency range decrease the final estimation performance.

The wideband MUSIC is typically successful in the whole SNR range, as well as when the signals are well separated from one to another, but it suffers from errors and produces side peaks at incorrect angles when the noise is high (SNR is low) and the targets are closely placed, which is common in many cases. Furthermore, the noise level is expected to be flat across the frequency range, which is not the normal condition [20].

3.4 Wideband Capon

The purpose of broadband Capon approach is to reduce the overall output power (through various frequency ranges) of a group of narrowband beamformers $\omega(f_k,\theta)$, $k = 1, \ldots, K$; while applying a collection of narrowband distortionless response requirements $\omega^H(f_k,\theta)a(f_k,\theta)$. The problem can be mathematically described as:

$$\min_{\omega(f_k,\theta)} Q(\theta) = \sum_{k=1}^{K} \omega^H(f_k,\theta)R_{xx}(f_k)\omega(f_k,\theta), \tag{21}$$

under the constraints:

$$\omega^H(f_k,\theta)a(f_k,\theta) = 1, \qquad \forall k \in [1,K] \tag{22}$$

This reduction issue is readily addressed utilizing the Lagrange multiplier approach. The minimizers $\omega(f_k,\theta)$ are narrowband Capon beamformers [21] and computed by:

$$\omega(f_k,\theta) = \frac{R_{xx}^{-1}(f_k)a(f_k,\theta)}{a^H(f_k,\theta)R_{xx}^{-1}(f_k)a(f_k,\theta)}, \qquad \forall k \in [1,K] \tag{23}$$

The broadband Capon spectra is given by [21]:

$$Q(\theta) = \sum_{k=1}^{K} q(f_k,\theta) = \sum_{k=1}^{K} \frac{1}{a^H(f_k,\theta)R_{xx}^{-1}(f_k)a(f_k,\theta)} \tag{24}$$

where:

$$q(f_k,\theta) = \frac{1}{a^H(f_k,\theta)R_{xx}^{-1}(f_k)a(f_k,\theta)}, \qquad k = 1, \ldots, K, \tag{25}$$

are the output powers of the narrowband MVDR beamformers $\omega(f_k,\theta)$. The wideband Capon spectra may now be generated by taking an average of the narrowband Capon outputs $q(f_k,\theta)$ over multiple frequency ranges.

4 Simulation and Discussion

In this part, we demonstrate several computer simulations to verify the effectiveness of the MUSIC and Capon methods for narrowband and wideband signals.
In the following examples, we use a ULA of $N = \{8, 10, 12\}$ antenna elements
equally separated by half of the wavelength λ, and $K = 500$ snapshots. Considering three identical energy sources that arise from $\{10°, 15°, 30°\}$ intruding on
the array.

The root mean square error (RMSE) of angle estimation is defined as:

$$\text{RMSE} = \frac{1}{M} \sum_{m=1}^{M} \sqrt{\frac{1}{100} \sum_{n=1}^{100} \left(\hat{\theta}_m(n) - \theta_m\right)^2} \tag{26}$$

where $\hat{\theta}_m(n)$ is the estimated value of θ_m for the nth Monte Carlo treatment,
and M denotes the number of sources. Note that the RMSE was used to test
the estimation performance.

4.1 DOA Estimation for Narrowband Signals

The Fig. 3a illustrates the spectra of the narrowband MUSIC and the Fig. 3b
depicts the spatial spectrum of the narrowband Capon, where the SNR is fixed
at 0 dB.

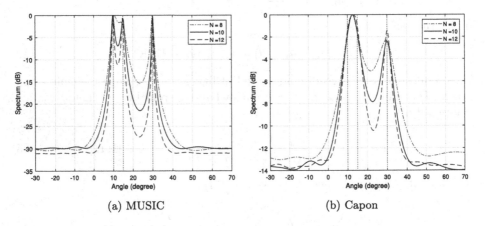

(a) MUSIC   (b) Capon

Fig. 3. Spatial spectrums of narrowband methods under different number of antennas
N, with $M = 3$ ($\theta_1 = 10°, \theta_2 = 15°, \theta_3 = 30°$), and SNR = 0 dB.

From Fig. 3a, we can deduce that when the number of antennas is wide, the
MUSIC algorithm is able to resolve the DOA estimation, which is 12 sensors.
When the number of antennas is small, for example 8, the narrowband MUSIC

cannot distinguish the two close signals. The Fig. 3b demonstrate that the narrowband Capon cannot detect accurately the two closest targets, and requires more than 12 antenna elements to estimate correctly the real DOAs.

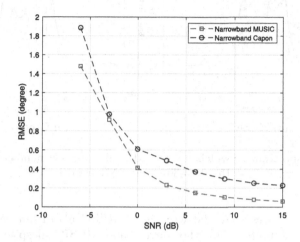

Fig. 4. RMSE results of narrowband methods versus SNR.

Table 1. RMSE values of narrowband methods versus SNR.

Method	SNR							
	−6 dB	−3 dB	0 dB	3 dB	6 dB	9 dB	12 dB	15 dB
MUSIC	1.479	0.918	0.413	0.233	0.151	0.105	0.078	0.060
Capon	1.886	0.975	0.610	0.490	0.372	0.299	0.252	0.228

The Fig. 4 and The Table 1 illustrate the RMSE results of the narrowband MUSIC against the narrowband Capon for $N = 8$ antenna elements and different SNRs varied from high-level noise (SNR $= -6$ dB) to low-level noise (SNR $= 15$ dB), by 3 steps. As depicted in Fig. 4 and Table 1, the narrowband MUSIC provides good performance more than the narrowband Capon method in the whole range of SNR. However, both algorithms suffer when the SNR is negative (SNR < 0 dB).

4.2   DOA Estimation for Wideband Signals

The Fig. 5a depicts the spectra of the wideband MUSIC and the Fig. 5b illustrates the spatial spectra of the wideband MVDR, where the value of SNR is 0 dB.

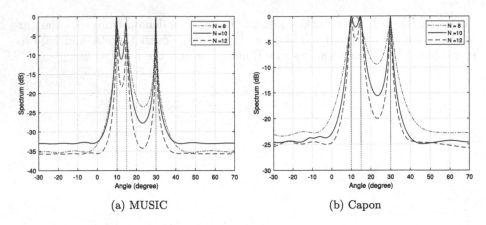

(a) MUSIC   (b) Capon

Fig. 5. Spatial spectrums of wideband methods under different number of antennas N, with $M = 3$ $(\theta_1 = 10°, \theta_2 = 15°, \theta_3 = 30°)$, and SNR $= 0$ dB.

The Fig. 5a proves that the wideband MUSIC algorithm can resolve the DOAs estimation better than the narrowband MUSIC, despite the fact that the number of antennas is minimal (8 sensors). When the number of antennas is large, the wideband MUSIC act better than its small. From Fig. 5b, we can see that the wideband Capon outperforms the narrowband Capon, and can estimate correctly the angles only with 10 elements.

Table 2. RMSE values of wideband methods versus SNR.

Method	\multicolumn{8}{c}{SNR}
	-6 dB	-3 dB	0 dB	3 dB	6 dB	9 dB	12 dB	15 dB
MUSIC	1.471	0.384	0.173	0.099	0.064	0.044	0.033	0.025
Capon	1.417	0.729	0.542	0.386	0.337	0.236	0.174	0.134

The Fig. 6 and the Table 2 illustrate the RMSE outcomes of the wideband MUSIC against the wideband Capon for $N = 8$ antenna elements and different SNRs varied from high-level noise (SNR $= -6$ dB) to low-level noise (SNR $= 15$ dB), by 3 steps.

The Fig. 6 and Table 2 show that the wideband MUSIC outperforms the wideband Capon in the whole range of SNR. However, both methods suffer when the SNR is negative (SNR < 0 dB).

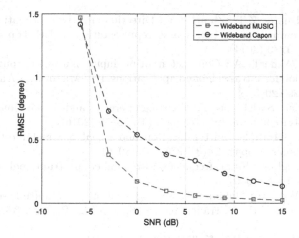

Fig. 6. RMSE results of wideband methods versus SNR.

5   Conclusion

In this paper, we have been presented an efficient comparative study between MUSIC and Capon approaches for 1D narrowband and wideband DOA estimation, in stationary case for ULA configuration, and for an additive white Gaussian noise. In order to evaluate their performance three arriving signals with considering that two of the signals are distanced by 5°, three different number of antennas were the key factor for evaluation. The result showed that the wideband version of these algorithms outperforms the narrowband version in high-level noise, in particular, MUSIC is more superior than Capon method. For the two nearest sources, the wideband MUSIC can accurately resolve the DOAs in high-level noise, but the narrowband MUSIC cannot locate them and works better in low-level noise.

The authors will attempt to analyze the partial covariance matrix approach in the situation of two-dimensional (2D) for broadband emitters in the future research.

References

1. Wu, X., Zhu, W.P., Yan, J.: A high-resolution DOA estimation method with a family of nonconvex penalties. IEEE Trans. Veh. Technol. **67**(6), 4925–4938 (2018)
2. Ebihara, S., Kimura, Y., Shimomura, T.: Coaxial-fed circular dipole array antenna with ferrite loading for thin directional borehole radar sonde. IEEE Trans. Geosci. Remote Sens. **53**(4), 1842–1854 (2015)
3. Nielsen, U., Dall, J.: Direction-of-arrival estimation for radar ice sounding surface clutter suppression. IEEE Trans. Geosci. Remote Sens. **53**(9), 5170–5179 (2015)
4. Takahashi, R., Inaba, T., Takahashi, T.: Digital monopulse beamforming for achieving the CRLB for angle accuracy. IEEE Trans. Aerosp. Electron. Syst. **54**(1), 315–323 (2018)

5. Liu, Z.M., Huang, Z.T., Zhou, Y.Y.: Direction-of-arrival estimation of wideband signals via covariance matrix sparse representation. IEEE Trans. Sig. Process. **59**(9), 4256–4270 (2011)
6. Toktas, A., Akdagli, A.: Compact multiple-input multiple-output antenna with low correlation for ultra-wideband applications. IET Microwaves Antennas Propag. **9**(8), 822–829 (2015)
7. Frikel, M., Safi, S., Khmou, Y.: Focusing operators and tracking moving wideband sources. J. Telecommun. Inf. Technol. (4), 53–59 (2016)
8. Schmidt, R.: Multiple emitter location and signal parameter estimation. IEEE Trans. Antennas Propag. **34**(3), 276–280 (1986)
9. Capon, J.: High-resolution frequency-wavenumber spectrum analysis. Proc. IEEE **57**(8), 1408–1418 (1969)
10. Ziskind, I., Wax, M.: Maximum likelihood localization of multiple sources by alternating projection. IEEE Trans. Acoust. Speech Sig. Process. **36**(10), 1553–1560 (1988)
11. Ottersten, B., Viberg, M., Kailath, T.: Analysis of subspace fitting and ML techniques for parameter estimation from sensor array data. IEEE Trans. Sig. Process. **40**(3), 590–600 (1992)
12. Aounallah, N.: Performance enhancement of Capon's DOA algorithm using covariance matrix decomposition. Eng. Proc. **14**(1), 7 (2022)
13. Liu, D., Li, Z., Guo, X., Zhao, S.: DOA estimation for wideband LFM signals with a few snapshots. EURASIP J. Wirel. Commun. Netw. **2017**(1), 1–7 (2017)
14. Ali, E., Ismail, M., Nordin, R., Abdulah, N.F.: Beamforming techniques for massive MIMO systems in 5G: overview, classification, and trends for future research. Front. Inf. Technol. Electron. Eng. **18**(6), 753–772 (2017). https://doi.org/10.1631/FITEE.1601817
15. Liu, Y., Chai, J., Zhang, Y., Liu, Z., Jin, M., Qiu, T.: Low-complexity neural network based DOA estimation for wideband signals in massive MIMO systems. AEU-Int. J. Electron. Commun. **138**, 153853 (2021)
16. Xu, Z., Wu, S., Yu, Z., Guang, X.: A robust direction of arrival estimation method for uniform circular array. Sensors **19**(20), 4427 (2019)
17. Chung, P.J., Bohme, J.F., Mecklenbrauker, C.F., Hero, A.O.: Detection of the number of signals using the Benjamini-Hochberg procedure. IEEE Trans. Sig. Process. **55**(6), 2497–2508 (2007)
18. Wax, M., Kailath, T.: Detection of signals by information theoretic criteria. IEEE Trans. Acoust. Speech Sig. Process. **33**(2), 387–392 (1985)
19. Hayashi, H., Ohtsuki, T.: DOA estimation for wideband signals based on weighted Squared TOPS. EURASIP J. Wirel. Commun. Netw. **2016**(1), 1–12 (2016). https://doi.org/10.1186/s13638-016-0743-9
20. Yoon, Y.S., Kaplan, L.M., McClellan, J.H.: TOPS: new DOA estimator for wideband signals. IEEE Trans. Sig. Process. **54**(6), 1977–1989 (2006)
21. Azimi-Sadjadi, M.R., Pezeshki, A., Roseveare, N.: Wideband DOA estimation algorithms for multiple moving sources using unattended acoustic sensors. IEEE Trans. Aerosp. Electron. Syst. **44**(4), 1585–1599 (2008)

A Novel Hybrid Approach for Improving the Accuracy of the Supervised Link Prediction Based on Graph Structure Features in Social Networks

Mohamed Badiy[1]([⊠])(iD), Fatima Amounas[2](iD), and Moha Hajar[3](iD)

[1] RO.AL&I Group, Faculty of Sciences and Technics, Moulay Ismaïl University,
Errachidia, Morocco
badaymohamed1995@gmail.com
[2] R.O.AL&I Group, Computer Sciences Department, Faculty of Sciences
and Technics, Moulay Ismaïl University, Errachidia, Morocco
[3] R.O.AL&I Group, Mathematical Department, Faculty of Sciences and Technics,
Moulay Ismaïl University, Errachidia, Morocco

Abstract. Now a day's, the analysis of social networks has grown significantly. Link prediction is a challenging issue in the social network analysis area, which uses existing network information in order to predict the future link. Many different link prediction techniques have been proposed in order to predict the future link in the network. Recently, supervised link prediction has become a growing field in which several research efforts have been made. In this context, this paper deals with a new hybrid approach to supervised link prediction based on the merging of local and global similarity methods. It attempts to improve the performance of supervised link prediction by combining various similarity measures. This research considers multiple topological features for training supervised machine learning classifiers. Practical implementation proved that the proposed approach revealed satisfactory results as compared to the existing methods. Our results show that using both local and global features outperforms similarity features when applied individually. Further, the hybridization of multiple features can achieve the highest accuracy.

Keywords: Social network · Link prediction · Feature extraction · Hybrid model · Supervised learning

1 Introduction

During the last few years, the field of social networks Analysis (SNA) has gathered special interest from research communities. The social network has become a platform to share data, mainly including text, images, videos, and so on, and

© Springer Nature Switzerland AG 2022
M. Fakir et al. (Eds.): CBI 2022, LNBIP 449, pp. 231–242, 2022.
https://doi.org/10.1007/978-3-031-06458-6_19

to start creating a number of connections between the entities. The connections provide a flexible way to describe the relationships between the users. The social network can be visualized as graph between multiple nodes [1]. In such graphs the nodes represent entities, e.g. people or organizations, linked together based on different factors, e.g. friendship. However, social networks have a dynamic structure, as new edges and vertices are added to the graph over time. Social Network Analysis is a large field of research that deals with the study and analysis of social networks. Link prediction is one of the most important task in social network analysis. This task deals with the problem of predicting potential future relationships between nodes based on observations in the current nodes and their relations. Link prediction has many applications in different areas. For example, in online social networks, link prediction helps users to find new friends [2]. In biology, it can be used to identify new interactions between genes, diseases, and drugs within interaction networks [3]. In spam mail detection, it can be used to detect anomalies in the emails [4]. It can also be used in a recommendation system for electronic commerce [5,6]. Due to the wide range of applications in different areas, link prediction problem has attracted significant research attention. Many link prediction approaches have been proposed to solve this problem. These approaches can be classified into two categories: similarity-based method and learning-based method. The similarity-based method includes four classes: local approaches, global approaches, quasi-local approaches, and community based approaches. These methods assign the similarity scores to the node pairs according to the structure of the networks [7,8]. The learning-based method considers the link prediction as a classification problem [9,10]. The learning-based method develops models based on training data in order to observe the new link being predicted. These methods produce better performance statistics than similarity-based methods. The feature selection is an important step in machine learning process. For this purpose, the network topology data is transformed into features that could be used in a predictive model. In the link prediction, the approaches mainly used for feature extraction are based on similarity, probabilistic and dimensionality reduction [11]. Currently, the supervised link prediction is one of the emerging research directions in the social network analysis. In this context, this research work aims to develop a hybrid approach to improve the accuracy of the supervised link prediction algorithms by using the merging of multiple topological metrics. The main goal of this work is to extract the feature vectors from the Facebook page and the Dolphin networks.

The paper is organized as follows: in the Sect. 2, we review the related works on link prediction. In Sect. 3, we give some basic theories connected with link prediction approaches. Section 4 is devoted to the proposed approach, followed by the experimental results and the performance evaluation. Finally, Sect. 5 concludes the paper.

2   Related Work

In the last years, the link prediction problem has been intensively studied by scientific community. Several link prediction algorithms using similarity-based methods have been proposed [12–14], but still there is a scope to improve the previous approaches. Recently, Machine learning has extensively contributed in the development of several link prediction methods. For instance, Ahmed et al. [15] introduced a supervised learning approach to predict missing links in Twitter. To achieve this goal, they have adopted various classifiers like Bagging, Decision Trees, Rotation Forest, with similarity indices as the features vectors. In [16], the authors investigated the link prediction problem in multiplex networks (Twitter and Foursquare). For the classification task on two-layer social networks, the authors used two sets of features: meta-path-based and node-based features. The experimental results indicate that the accuracy score improved by including the cross-layer information. Another similar work is done by Fu et al. [9] in weighted networks. The authors put forward a supervised learning approach based on similarity and centrality indices to perform the link weight prediction task.

In [17], a new supervised learning method is given to solve the link prediction problem in two-layer networks. They show that the prediction of links in the Foursquare network achieves better accuracy than the Twitter network. Another important approach to be mentioned is given by Bütün et al. [18]. Here, the authors proposed a link prediction by considering both the link directions and the topological patterns. Hard datasets are generated in order to improve the link prediction performance of the Triad Closeness (TC) metric in HEP-Th and DBLP networks. The results show that the proposed approach achieves the highest accuracy in the TC metric as compared to the state-of-the-art link prediction metrics. Shan et al. [19] studied the link prediction problem in multiplex networks. The authors developed a novel approach which treats link prediction as a binary classification problem. In the proposed method, the goal is to predict the labels of the node pairs that are extracted from different layers. The results show that the proposed method outperforms the other methods designed in a single layer. Another work has been carried out in this field by Gerrit Jan de Bruin et al. [20]. The authors addressed the supervised temporal link prediction on 26 temporal networks. They prove that the temporal activity of the node or the edge hat in impact on the link prediction performance.

More recently, Malhotra Deepanshu and Rinkaj Goyal [10] put forward a solution for predicting links in single-layer and multiplex networks. They adopted a supervised machine learning technique using topological features of the network for training the models. The results show the high accuracy by testing the proposed method on various real-world network data sets. Further, Ghorbanzadeh, Hossien, et al. [21] proposed a novel measure based on common neighborhood for the link prediction on directed networks. The result shows that their measure exhibited the best performance in unsupervised and supervised learning modes. In this context, the present paper attempts to propose the supervised link prediction approach with improved structural features.

3   Preliminaries

In this section, we give an overview of certain topics. This will serve as a background and introduction of the concepts which are used in the proposed approach.

3.1   Link Prediction Problem in Graph

Link prediction is a fundamental problem in graph data analysis. Link prediction problem in a graph can be defined as a classification problem, where non-existing links are classified as Predicted (P) or Not Predicted (NP). Let G (V, E) be a graph of social network where V denotes set of users and E denotes set of links that connects two users in the network. Our goal is to find the label of the non-existing links.

Let L be the set of labels for the non-existing links (either P or NP). We formally define a new prediction function, which helps to find the label of the non-existing links. Mathematically, it can be formulated as follows:

$$f : V \times V \to L$$
$$(v_x, v_y) \to P/NP$$

3.2   Supervised Machine Learning Methods Adopted

Machine supervised learning is one of the main paradigms of link prediction. In this section, we briefly review the supervised machine learning techniques that we use in our experiments.

- *Decision Tree (DT)*: is one of the most popular machine learning algorithms for classification and regression problems, but mostly it is preferred for solving classification problems [22]. Decision tree are trees that classify instances by sorting them based on feature values, where each node represents the features of a dataset and each branch represents a value that the node can assume.
- *Logistic Regression (LR)*: is another simple supervised machine learning classifier used to predict the probability of a binary event occurring such as 0 and 1, true or false, negative or positive, and no or yes [23]. Logistic regression describes and estimates the relationship between one dependent binary variable and one or more independent variables.
- *Naive Bayes (NB)*: is a probabilistic machine learning model that's used for classification task in which instances of the dataset are discriminated based on the specified feature. It is based on the Bayes' theorem with the assumption of independence between each pair of features [24].
- *XGBoost*: also well-known by extreme Gradient Boosting is an extention of the gradient boosted decision trees [25], which is known as one of the best performing algorithms utilized for supervised learning. It can be used for both regression and classification problems. It is used for wide number of applications and it also supports outdoor memory. Due to parallel computation process, it is faster than other boosting algorithms. The reason behind the higher performance of XGBoost is that it is scalable in nature.

3.3   Feature Selection

Over the past decades, supervised link prediction has attracted extensive attention of most researches. The feature extraction an important step in the supervised link prediction task. According to the link prediction taxonomy [11], the feature extraction techniques can be classified under two main categories: feature-extraction and feature-learning techniques. The most commonly used features are "feature extraction methods." Figure 1 illustrates the different feature extraction techniques of link prediction in social networks.

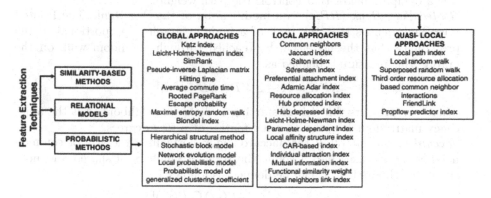

Fig. 1. A taxonomy for the feature extraction techniques in link prediction task.

Recently, many studies have examined the effectiveness of different feature selection algorithms based on different data sets. In this context, we consider some topological features as the features vectors for training our models. More precisely, we evaluate the effectiveness of supervised learning methods using the combination of six topological features including Shortest Path (SP), Katz measure, Rooted Page Rank (RPR), Jaccard coefficient (JC), Preferential attachment (PA) and Resource allocation Index (RA). Some of the notations related to feature metrics are discussed as below:

- $\Gamma(v_x)$: neighbors set of node v_x.
- $\Gamma(v_y)$: neighbors set of node v_y.
- $|\Gamma(v_x)|$: number of neighbors of node v_x.
- $|\Gamma(v_y)|$: number of neighbors of node v_y.

The different similarity features that we have adopted in our work are as follows:

- *Shortest Path (SP)*: this measure calculates the similarity of two nodes v_x and v_y by considering the shortest path between them and reflecting the expectation that nodes with low degrees of separation are likely to interact. This measure can be calculated as follows:

$$SP(v_x, v_y) = \min(|Pv_x \rightarrow v_y|)$$

where $Pv_x \rightarrow v_y$ is a path between v_x and v_y, whereas, $|P|$ denotes the length of path P.

- *Katz measure*: the katz index determined the number of paths between two nodes and discounts the longer path exponentially. It can be considered as a variant of the shortest path metric. It can be computed as:

$$Katz(v_x, v_y) = \sum_{k=1}^{\infty} \beta^k \cdot |paths_{v_x,v_y}^k|$$

where $paths_{v_x,v_y}^k$ is a collection of all paths with length k between v_x and v_y. β is a damping factor that controls the path weights.

- *Rooted Page Rank (RPR)*: is a modification of the PageRank. The Rooted PageRank defines that the rank of a node in graph is proportional to the probability that the node will be reached through a random walk on the graph. This measure is defined as:

$$RPR = (1 - \beta)[(I - \beta D^{(-1)}A)]^{(-1)}$$

where β represents the visit of starting node to its neighbors, A is the adjacency matrix and D is the diagonal matrix of vertex degrees.

- *Jaccard coefficient* (JC): the Jaccard index normalizes the size of common neighbors and takes into account the total of numbers of shared and non-shared neighbors [26]. It can be defined as follows:

$$JC(v_x, v_y) = \frac{|\Gamma(v_x) \bigcap \Gamma(v_y)|}{|\Gamma(v_x) \bigcup \Gamma(v_y)|}$$

- *Preferential attachment* (PA): the preferential attachment index calculates the similarity score for each pair of nodes using their degree information. It can be computed as below:

$$PA(v_x, v_y) = |\Gamma(v_x)| \cdot |\Gamma(v_y)|$$

- *Resource allocation Index* (RA): this metric is very similar to adamic and adar index. The Resource Allocation uses the degree magnitude instead of logarithmic value [27]. This measure is calculated as follows:

$$RA(v_x, v_y) = \sum_{k \in |\Gamma(v_x) \bigcap \Gamma(v_y)|} \frac{1}{|\Gamma(k)|}$$

4   Methodology

In this paper, we present a supervised approach with some features vectors to solve link prediction problem in social network. We also evaluate the effectiveness of supervised learning methods using different features vectors for link prediction in social network. Our goal is to see how effective these features are in terms of predicting the future links in a social network. Now, we discuss the proposed approach in greater detail to explain its working. Here, we use four well-known supervised machine learning methods as discussed in Sect. 3.2. The proposed architecture of the link prediction process using the supervised learning algorithms is described in Fig. 2.

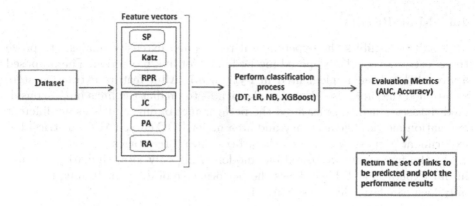

Fig. 2. An overview diagram of the proposed approach

4.1   Benchmark Datasets and Settings

In this paper, we use two multiplex networks Facebook Pages and Dolphin to
verify the effectiveness of proposed models. Facebook Pages [28] represents is an
unweighted and undirected social network. In this network, the pages are the
nodes, and an edge is represented by an associated link between two nodes. The
Dolphin dataset [29] is a social network of bottlenose dolphins. In this network,
the bottlenose dolphins are represented by nodes and an edge is an association
between two nodes. The detailed statistics are summarized in Table 1.

Table 1. Data sets considered.

Network	Nodes	Edges	Average degree	Clustering coefficient	Density
Facebook pages	620	2102	6	0.330897	0.0108969
Dolphin	62	159	5	0.258958	0.0840825

4.2   Feature Extraction Process

In supervised link prediction, it is very important to choose the feature vectors
in order to build the classifiers. In this work, the feature vector is constructed by
considering both local or global information in the network. As discussed above,
for the feature selection process, a subset of features is selected from the original
feature set {JC; PA; RA; SP; Katz; RPR}. The first three, that is, the Jaccard
coefficient, preferential attachment, and resource allocation index, represented
the local similarity measures. Furthermore, Shortest Path, Katz measure, and
Rooted Page Rank represented the global similarity measures. As many features
are considered, the extraction feature process is categorized into three stages.
The first one is based on the local similarity features (Stage 1). The second one
is based on global similarity features (Stage 2). In the last one, the local feature
are combined with the global features (Stage 3).

4.3   Main Results

This section exhibits the experimental results and extensive analysis to prove the effectiveness of the proposed methodology for link prediction. The proposed approach has been implemented in Python 3.7. After feature extraction step, we applied the supervised learning algorithms to predict the links in the considered dataset. Then, we measured the performance of our models using different evaluation metrics like accuracy and area under ROC curve (AUC) metrics. For experiment purpose, we divided the process into three stages.

In the first stage, we tested our models using only local similarity features in our experiment. Table 2 shows the performance of different techniques using only local similarity features (Stage 1).

Table 2. Performance of the models using only the local similarity features

Dataset	Classifier	AUC	Accuracy
Facebook page	DT	0.516	0.927
	LR	0.492	0.927
	NB	0.513	0.90
	XGBoost	0.513	0.927
Dolphin	DT	0.497	0.948
	LR	0.529	0.948
	NB	0.543	0.921
	XGBoost	0.550	0.948

From the results obtained, it can be seen that local methods cannot achieve satisfactory performance on all evaluation metrics.

Next, we investigated the performance of the models using only global similarity features in the second stage. Table 3 shows the performance of different techniques using only global similarity features (Stage 2).

Table 3. Performance of the models using only the global similarity features

Dataset	Classifier	AUC	Accuracy
Facebook page	DT	0.742	0.949
	LR	0.936	0.932
	NB	0.939	0.921
	XGBoost	0.971	0.951
Dolphin	DT	0.654	0.948
	LR	0.800	0.948
	NB	0.823	0.941
	XGBoost	0.910	0.948

From the above results, it observed that all the models, except DT model have a performance greater than 80% in term of AUC, and a performance greater than 90% in term of accuracy.

Finally, we investigated the performance of different models using the combination of local and global features in the stage 3. The Table 4 illustrates the obtained results by considering multiple features (Stage 3).

Table 4. Performance of the models using both the local and global similarity features

Dataset	Classifier	AUC	Accuracy
Facebook page	DT	0.808	0.950
	LR	0.940	0.964
	NB	0.942	0.945
	XGBoost	0.974	0.951
Dolphin	DT	0.721	0.974
	LR	0.804	0.970
	NB	0.881	0.952
	XGBoost	0.989	0.981

In the current case, the accuracy of the models are very close, and better than using the features separately. It can be observed that all the models have a performance greater than 94% in term of accuracy. Our results show that the XGBoost model achieved 98% in all metrics, and DT model has the lowest AUC values (0.808 and 0.721). The graphical representation of the above described tables is shown in Fig. 3, Fig. 4 and Fig. 5 (respectively).

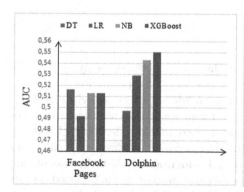
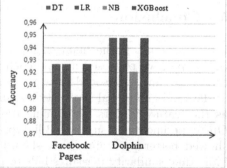

Fig. 3. Performance analysis of different link prediction models using the local similarity features

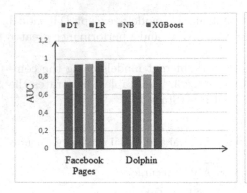
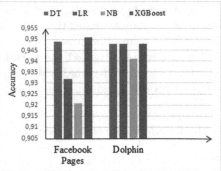

Fig. 4. Performance analysis of different link prediction models using the global similarity features

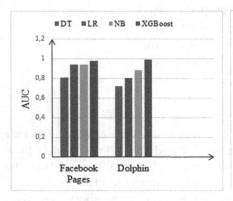
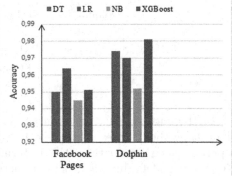

Fig. 5. Performance analysis of different link prediction models using the combination of local and global similarity features

5   Conclusion

Link prediction is the main challenge in the social network analysis field. In this paper, we proposed an approach of supervised link prediction to predict the future link in a social network. Here, we made an effort to hybridize both of the local and global similarity measures. Then, we considered these measures as the feature vectors for training the supervised machine learning classifiers. Our experiments on supervised link prediction revealed satisfactory results and showed better performance. It should also be mentioned that the use of any individual similarity measures fails to predict the in-existent link. Based on our experiments, we observed that the combination of different similarity measures is a very effective feature for supervised link prediction. Finally, we would like to point out that the hybridization of multiple features will provide better performance in this regard. As a future work, the proposed approach can be extended to link prediction in big social networks.

References

1. Liben-Nowell, D., Kleinberg, J.: The link-prediction problem for social networks. J. Am. Soc. Inf. Sci. Technol. **58**, 1019–1031 (2007). https://doi.org/10.1002/asi.20591
2. Aiello, L.M., et al.: Friendship prediction and homophily in social media. ACM Trans. Web (TWEB) **6**(2), 1–33 (2012). https://doi.org/10.1145/2180861.2180866
3. Lin, C.-H., et al.: Multimodal network diffusion predicts future disease-gene-chemical associations. Bioinformatics **35**, 1536–1543 (2019). https://doi.org/10.1093/bioinformatics/bty858
4. Huang, Z., Zeng, D.D.: A link prediction approach to anomalous email detection. In: 2006 IEEE International Conference on Systems, Man and Cybernetics, vol. 2, pp. 1131–1136. IEEE (2006). https://doi.org/10.1109/ICSMC.2006.384552
5. Akcora, C.G., Carminati, B., Ferrari, E.: Network and profile based measures for user similarities on social networks. In: 2011 IEEE International Conference on Information Reuse & Integration, pp. 292–298. IEEE (2011). https://doi.org/10.1109/IRI.2011.6009562
6. Wu, S., Sun, J., Tang, J.: Patent partner recommendation in enterprise social networks. In: Proceedings of the Sixth ACM International Conference on Web Search and Data Mining, pp. 43–52 (2013). https://doi.org/10.1145/2433396.2433404
7. Hou, L., Liu, K.: Common neighbour structure and similarity intensity in complex networks. Phys. Lett. A **381**, 3377–3383 (2017). https://doi.org/10.1016/j.physleta.2017.08.050
8. Muniz, C.P., Goldschmidt, R., Choren, R.: Combining contextual, temporal and topological information for unsupervised link prediction in social networks. Knowl. Based Syst. **156**, 129–137 (2018). https://doi.org/10.1016/j.knosys.2018.05.027
9. Fu, C., et al.: Link weight prediction using supervised learning methods and its application to yelp layered network. IEEE Trans. Knowl. Data Eng. **30**, 1507–1518 (2018). https://doi.org/10.1109/TKDE.2018.2801854
10. Malhotra, D., Goyal, R.: Supervised-learning link prediction in single layer and multiplex networks. Mach. Learn. Appl. **6**, 100086 (2021). https://doi.org/10.1016/j.mlwa.2021.100086
11. Daud, N.N., et al.: Applications of link prediction in social networks: a review. J. Netw. Comput. Appl. **166**, 102716 (2020). https://doi.org/10.1016/j.jnca.2020.102716
12. Yu, C., Zhao, X., An, L., Lin, X.: Similarity-based link prediction in social networks: a path and node combined approach. J. Inf. Sci. **43**(5), 683–695 (2016). https://doi.org/10.1177/0165551516664039
13. Zhou, K., Michalak, T.P., Waniek, M., Rahwan, T., Vorobeychik, Y.: Attacking similarity-based link prediction in social networks. In: Proceedings of the 18th International Conference on Autonomous Agents and Multi Agent Systems, pp. 3, 5–313 (2019). https://doi.org/10.48550/arXiv.1809.08368
14. Kumari, A., Behera, R.K., Sahoo, K.S., Nayyar, A., Kumar Luhach, A., Prakash Sahoo, S.: Supervised link prediction using structured based feature extraction in social network. Concurr. Comput. Pract. Exp. e5839 (2020). https://doi.org/10.1002/cpe.5839
15. Ahmed, C., ElKorany, A., Bahgat, R.: A supervised learning approach to link prediction in Twitter. Soc. Netw. Anal. Min. **6**(1), 1–11 (2016). https://doi.org/10.1007/s13278-016-0333-1

16. Jalili, M., et al.: Link prediction in multiplex online social networks. Roy. Soc. Open Sci. **4**, 160863 (2017). https://doi.org/10.1098/rsos.160863
17. Mandal, H., et al.: Multilayer link prediction in online social networks. In: 2018 26th Telecommunications Forum (TELFOR), pp. 1–4. IEEE (2018). https://doi.org/10.1109/TELFOR.2018.8612122
18. Ertan, B., Kaya, M.: A pattern based supervised link prediction in directed complex networks. Phys. A **525**, 1136–1145 (2019). https://doi.org/10.1016/j.physa.2019.04.015
19. Shan, N., et al.: Supervised link prediction in multiplex networks. Knowl. Based Syst. **203**, 106168 (2020). https://doi.org/10.1016/j.knosys.2020.106168
20. de Bruin, G.J., Veenman, C.J., van den Herik, H.J., Takes, F.W.: Supervised temporal link prediction in large-scale real-world networks. Soc. Netw. Anal. Min. **11**(1), 1–16 (2021). https://doi.org/10.1007/s13278-021-00787-3
21. Ghorbanzadeh, H., et al.: A hybrid method of link prediction in directed graphs. Expert Syst. Appl. **165**, 113896 (2021). https://doi.org/10.1016/j.eswa.2020.113896
22. Muhammad, I., Yan, Z.: Supervised machine learning approaches: a survey. ICTACT J. Soft Comput. **5** (2015). https://doi.org/10.21917/ijsc.2015.0133
23. Ong, E., et al.: COVID-19 coronavirus vaccine design using reverse vaccinology and machine learning. Front. Immunol. 1581 (2020). https://doi.org/10.1101/2020.03.20.000141
24. Saritas, M.M., Yasar, A.: Performance analysis of ANN and Naive Bayes classification algorithm for data classification. Int. J. Intell. Syst. Appl. Eng. **7**, 88–91 (2019). https://doi.org/10.18201//ijisae.2019252786
25. Chen, T., Guestrin, C.: XGBoost: a scalable tree boosting system. In: International Conference on Knowledge Discovery Data Mining, pp. 785–794 (2016). https://doi.org/10.1145/2939672.2939785
26. Jaccard, P.: Étude comparative de la distribution florale dans une portion des Alpes et des Jura. Bull. Soc. Vaudoise. Sci. Nat. **37**, 547–579 (1901). https://doi.org/10.5169/seals-266450
27. Tao, Z., Linyuan, L., Yi-Cheng, Z.: Predicting missing links via local information. Eur. Phys. J. B Condens. Matter Complex Syst. **71**, 623–630 (2009). https://doi.org/10.1140/epjb/e2009-00335-8
28. Rossi, R., Ahmed, N.: The network data repository with interactive graph analytics and visualization. In: Twenty-Ninth AAAI Conference on Artificial Intelligence (2015)
29. Lusseau, D., Schneider, K., Boisseau, O.J., et al.: The bottlenose dolphin community of Doubtful Sound features a large proportion of long-lasting associations. Behav. Ecol. Sociobiol. **54**, 396–405 (2003). https://doi.org/10.1007/s00265-003-0651-y

Intelligent System Based on GAN Model for Decision Support in Brain Tumor Segmentation

Omar El Mansouri[1]([⊠]) [iD], Yousef El Mourabit[1] [iD], Youssef El Habouz[2],
Nassiri Boujemaa[3], and Mohamed Ouriha[1] [iD]

[1] TIAD Laboratory, Sciences and Technology Faculty, Sultan Moulay Slimane University,
Beni Mellal, Morocco
omar.elmansouri@usms.ma
[2] IGDR UMR 6290 CNRS Rennes 1 University, Rennes, France
[3] Sustainable Innovation and Applied Research Laboratory, Polytechnique School, International
University of Agadir Morocco, Agadir, Morocco

Abstract. The most prevalent malignant brain tumors are gliomas, with a variety of grades, and each grade has a significant impact on a patient's chances of survival. Low-grade gliomas are usually found in the human brain and spinal cord. Low-grade glioma may be accurately diagnosed and detected early, lowering the risk of mortality for patients. In the examination gliomas of low grade, segmentation of MRI images is critical. The result, manual of Segmentation Techniques takes a long time and require a lot of pathology knowledge. in our study, we provide a unique generative adversarial network-based approach for segmenting images of tumors in the brain. The network is a structure between two neurons the generator and the discriminator. The generator is taught to construct an input mask of a take original image, The discriminator can tell the difference between the original and created masks, the end goal is to create masks for the input. The suggested model achieves a dice result of 0.97 in generalized experimental results from the TCGA LGG dataset, with a loss coefficient of 0.030, which is more effective and efficient than the compared approaches.

Keywords: TCGA-LGG · Deep learning · Segmentation · GAN

1 Introduction

Recent developments in vision computing and techniques for learning have improved proven to be very satisfying when applied to difficult, precise, and time-consuming tasks that may require specialized knowledge to perform when it comes to medical field imaging, these technologies can help medical specialists conduct delicate jobs and make critical and decisions. Medical imaging approaches using neural network models and deep learning algorithms have been developed, medical personnel now have access to advanced technology and effective tools to help their improvement their diagnosis [1]. Among medical imaging techniques, X-ray, CT, and MRI, these techniques consist of

© Springer Nature Switzerland AG 2022
M. Fakir et al. (Eds.): CBI 2022, LNBIP 449, pp. 243–253, 2022.
https://doi.org/10.1007/978-3-031-06458-6_20

two-dimensional or three-dimensional images to record vital and information that is useful from within the body to diagnostic purposes, planning treatment, and especially difficult and time-consuming decision making (i.e., patient diagnosis) [6]. Artificial intelligence, deep learning, and computer vision have become major components of doctors' expertise in two different areas: classification image and organ segmentation. Currently, as in domain of image segmentation for medical applications, medical pictures such as MRI are manually annotated, which is a time-consuming and somewhat subjective process. MRI is particularly useful for evaluating malignant brain tumors. Among the most common tumors, gliomas come in different grades, and each grade has a substantial influence just on patient's survival chances. Lower gliomas are most commonly detected in the human brain and spinal cord. Early identification and accurate diagnosis of minimal gliomas can lower the risk of mortality in afflicted individuals [6]. Thus, any progress toward successful automation of these tasks would greatly benefit both patients and medical experts. While certain cancers, such as gliomas, are simple to segment, others, such as gliomas tumors, are more difficult. Manual segmentation is resource-intensive and requires a high level of expertise from a pathologist. AI offers access to fresh knowledge and enhances diagnosis accuracy in support of physicians and their expertise. As a result, healthcare providers now have access to strong and effective technologies that can help them enhance their prognosis. With neural networks and deep learning technologies being used, IA is becoming a critical tool in the creation and medicine's advancement [1]. For this, we need to find an algorithm that can perform well in the medical imaging field for gliomas in terms of segmentation.

In terms of our challenge, we concentrated on a collection of research aiming at developing a segmentation algorithm. In recent years, artificial neural networks (CNNs) have been widely used to solve image recognition difficulties, and They've also been demonstrated to be useful in learning a hierarchy of qualities from several layers of data. CNNs have also been successful in per-pixel semantic segmentation. For example, Long et al. introduced semantic segmentation in their early work on integer convolutional networks (FCNs) in [2]. The authors create a coarse label map by CNNs often use fully linked layers with convolutional layers, the label map is then tested with deconvolutional layers to provide classification results with each pixel. The authors [3] presented the U-net, a U-shaped structure that consists of two sections: encoder and decoder, for the job of segmenting brain structures in electron microscopy. To increase hashing performance, they used jump connections between the encrypted layers and decryption. In [4] describes a network and learning strategy that employs an architecture that includes a shrinkage path for context capture and an asymmetric augmentation path for correct localization, and depends on the reliable data augmentation is used to make greater use of existing samples with annotations. They show that this network can all be consistently trained reliably on a small number of images and that it outperforms the previous best approach (sliding convolutional network) in the ISBI task of segmenting neuronal structures in electron microscopic ensembles [5]. In this paper, the authors present the implementation for Mask-RCNN, To the field of oral pathology, a contemporary convolutional neural network approach for object detection and segmentation. Mask-RCNN was designed with the goal of recognizing and segmenting object instances in natural images. They have demonstrated that Mask-RCNN can be employed in a highly specialized sector such as oral illnesses as a result of this experience. The authors of [7] developed SegAN, a technique for segmenting brain images that makes use of a minimum-maximum game to train the segmentation and critic networks alternately: As input, the reviewer is given two photographs (original image, expected label map, image original, label map), The segment is then trained by maximizing the multiscale loss function; the critic travels only the gradients that the segment is trained on, with the objective of decreasing the multiscale loss function. Work has proven that the SegAN architecture is more efficient and reliable for hashing tasks than the contemporary U-net hashing approach, and that it provides greater performance. The author presented to use an adversarial network topology with a pay network to drive a convolution layers hash network in order to obtain high-resolution segmentation in thoracic Xray organs using the dataset JSRT, and they were successful in obtaining a competitive dice score that was better than modern segmentation methods (U-net, Mask-RCNN) [8]. Recently, the authors proposed [9] utilizing antagonistic training to improve semantic segmentation and training for enhancement using models like the VGG network, which classifies pictures into a scale. The authors used synthetic data to train a GAN method to search for lung nodes, and a PHNN for segmentation, yielding very substantial segmentation results [10]. The authors employed the JSRT dataset [12] to develop a well-generalized U-Net attention model for X-ray images the chest [11], which obtained a final DSC of 97.5% [21]. In this study, the authors present model of segmentation in chest xrays using GAN and JSRT dataset which obtained a final DSC of 97.78% [13]. In this study, the authors present a comprehensive automated model for segmenting brain cancer tumors using GANs. This technique combines the generative and characteristic models and employs GAN as a high-level smoothing method instead of random fields with conditions (CRFs). In comparison to recent segmentation algorithms, they generated respectable segmentation results. The adversarial network technique has been found to be capable of segmenting glioma medical imaging jobs through our assessment of a set of studies. As a result, we employed generative adversarial networks to segregate brain tumors in our work. This model comprises a generator that is similar to the U-net design, as well as a discriminator that is created and sorted across truth masks. We've demonstrated that our model generates an extremely realistic and accurate hash, with a die score of 0.97, which is more effective and useful than comparable approaches.

The rest of this work is arranged in the following. Introduces generative adversarial networks as well as our model's technique in Sect. 2. Then, in Sect. 3, look at the findings of the experiment. Finally, in Sect. 4, we bring this article to a conclusion.

2 Methodology

In this section, an overall architecture of a generative adversarial network is detailed in this part, together with the discriminant and generator architectures described in our previous work, as well as a description of the data we will use to test our model.

2.1 Generative Adversarial Network

Goodfellow et al. presented Generative Adversarial Networks (GAN) in 2014 as a machine learning framework for learning the fundamental probability distribution of a high-dimensional training set and generating fake instances from the same distribution that are not part of the original data set [14]. A GAN is built up of two neural networks, one for the generator and one for the discriminator. The generator's (G) job is to learn instances of output from its estimated distribution to capture the data's distribution, while the discriminator (D) is to estimate the probability that the output of the generator is derived from the (real) data used for training, instead of from G. The two networks compete by training them against each other in a two-player game, where the discriminator's purpose to be enhance the likelihood of correctly labeling both the training examples and the generator output. The generator's goal is to increase the likelihood that the discriminator will not hit. The GAN's architecture is seen in Fig. 1. By reducing and maximizing the V, the function of the original GAN is supplied, as detailed in [15], Eq. (1) shows that:

$$\min_{G} \max_{D} V(G, D) = \mathbb{E}_y\big[\log D(y)\big] + \mathbb{E}_z\big[\log G(z)\big]. \tag{1}$$

So, y real information, z represents random data, D(y) the discriminator's prediction on y, G(z) represents data created by z noise and D(G(z)) represents the discriminator's prediction on the generated data. Both the generator and the discriminators play the min-max game and modify their settings in reaction to each other's actions. After a number of learning iterations, it's possible that both networks have parameters that can't be improved optimized. The generator now creates synthetic data that appears to be genuinely realistic, and the discriminator is unable to separate the real data from the produced data.

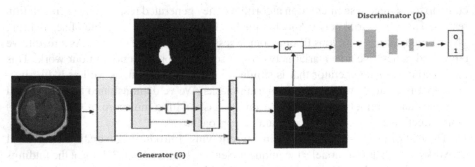

Fig. 1. Our framework's architecture. G the generator's structure is a U-net network to which we feed the X-ray images and their mask. The D discriminating network is a basic convolutional neural network that uses the ground truth picture to classify the image created by the generator into binary, such as "0" or "1."

2.2   Generator

Since GANs can be considered as fairly high-level standard optimizations, we need the generative part to improve the raw result obtained by the hash part. In our work, we used the generator which is a U-Net neural network instead of frequently used approaches like Mask-RCNN, it has shown good results in prior studies. U-Net [4] is a segmentation neural network that was developed of biological images (specifically, detect cell boundaries in biomedical images), which receives an image as input and outputs a Labeling map for each pixel (or similar) called the segmentation mask). The U-Net is a network design that has been improved to improve segmentation in medical imaging. U-Net follows a so-called autoencoder structure composed of two main structures: an encoder containing layers of convolutional and maximum assembly whose purpose to be reduce the input to a set of low-dimensional abstract representations of the original input, then there's the decoder, which is made up consists of a set of convolutional layers. A single hot coding is the ultimate result., i.e., it is a multi-channel output with a binary value where the channels correspond to the different classes and the ones and zeros of each channel indicate the presence or absence of that particular class at certain coordinates. To avoid the loss of high-resolution data information obtained during the encoding process and to preserve details that may be lost when the input is shortened by convolution and grouping, so-called jump connections are placed between the convolution layers and the offset positions of the same shape. This is simply a matter of saving a copy of each coding layer before maximum stacking, and concatenating (i.e., stacking) the channel with decoding layers of the same shape, before performing the transform convolution, Fig. 2 shown the overall construction of the generator.

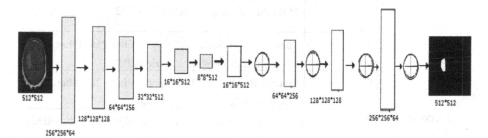

Fig. 2. Generator structure.

In the work, the generator captures two images, real image and a mask. Finally, it generates a segmented image with identical dimensions as the source image. This work's network is made by with six layers for the encoder and decoder, one is the inputs, four of the Relu convolutional batch normalization layers model is hidden, and the last layer is the output layer. In terms of the traits that were utilized, they are complete in Table 1.

The architecture consists of 12 layers. The encoder and decoder contain the six-layer each.

The encoder is the first half of the architecture composed of 6 layers. Each layer is composed as follows:

Input size is input data size, Conv2D is a two-dimensional convolution layer that has a convolution kernel wrapped in layers, allowing for the development of a tensor of output. Filters are masks that may be applied to images to blur them, sharpen and other effects convolving a filter and image, the pool_size layer is used to reduce representations and speed up calculations, as well as to make some of the features it detects little more toughness. Relu is indeed a piece-wise linear function that, if the input is positive,

Table 1. Architecture of the Generator in our model.

	Layer	Features
	1	Input size = (512,512,3), Conv2D, filter 64, Filter_size 3*3, pool_size 2*2, relu, padding='same'.
	2	Conv2D, filter 128, Filter_size 3*3, pool_size 2*2, relu, padding='same'.
	3	Conv2D, filter 256, Filter_size 3*3, pool_size 2*2, relu, padding='same'.
Encoder	**4**	Conv2D, filter 512, Filter_size 3*3, pool_size 2*2, relu, padding='same'.
	5	Conv2D, filter 512, Filter_size 3*3, relu, padding='same'.
	6	Conv2D, filter 512, Filter_size 3*3, pool_size 2*2, relu, padding='same'.
	1	Concatenate, Conv2DTranspose, filter 512, Filter size 3*3, relu, padding='same', Strides 2*2.
	2	Concatenate, Conv2DTranspose, filter 256, Filter size 3*3, relu, padding='same', Strides 2*2.
	3	Concatenate, Conv2DTranspose, filter 128, Filter size 3*3, relu, padding='same', Strides 2*2.
Decoder	4	Concatenate, Conv2DTranspose, filter 64, Filter size 3*3, relu, padding='same', Strides 2*2.
	5	Concatenate, Conv2DTranspose, filter 64, Filter size 3*3, relu, padding='same', Strides 2*2.
	6	Conv2D, filter 1, Filter size 1*1, sigmoid.

will directly output it. Otherwise, it will output zero, and padding is a Conv2D class parameter that can have one of two values: 'valid' means that the input volume is not completely filled with zero and Convolution can naturally reduce spatial dimensionality, or 'same' Convolution can automatically minimize the spatial dimensions because the input volume is not filled with zero.

The architecture's second half is the decoder is made up of six layers. Each compound layer is as follows:

Conv2DTranspose is a upsampling technique. After the transposed convolution, this image is then concatenated with the matching image from the contracting path, resulting in an image with the size input. The purpose of incorporating past data is to improve accuracy.

The finally is convolution layer with a filter size of 1 x 1 is the last layer. The activation sigmoid is defined on real numbers and its output is in the range (0, 1), which can be interpreted as a probability [17].

2.3 Discriminator

A discriminator a convolutional network that inserts two images, one real and one the generator's output. It then classifies each image as false or real. Figure 3 shown the overall structure of our discriminator.

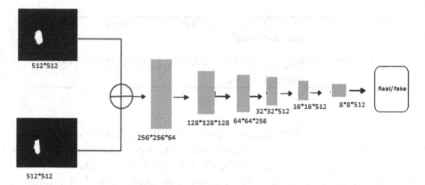

Fig. 3. Discriminator structure.

In this work, the discriminator combines two images, an original mask and a mask segmented by the generator, to create a sequence that can be classified as real or false. Binary entropy and loss dice were used be optimize the model. The network in this study has six layers: the first output layer is the input layer, four layers of the Relu convolution batch normalization model are hidden, and the last output layer is the final output layer. Table 2 lists all of the characteristics that were used:

2.4 Datasets

To assess our model, we used data from the Cancer Genome Atlas' low-grade glioma dataset (TCGA-LGG) for training and validation. This dataset is part of a wider effort to develop a research community focused on correlating cancer phenotypes to genetics by providing matching clinical image. The TCGA [16] topics. This dataset contains MR images of the brain along with manual masks to segment FLAIR abnormalities. They are 110 in number patients from the TCGA collection of low-grade gliomas for which at least FLAIR sequencing and genome mass data are available. Tumor genomic groups and patient data are provided in the "data.csv" file. Images were grouped in ".tif" format with three channels for each image. There are three sequences accessible for 101 cases: pre-contrast, FLAIR, and post-contrast. In 9 cases the postcontrast sequence is

Table 2. Architectures of the discrimination in our model.

Layer	Features
1	Input size = (512,512,3), Conv2D, filter 64, Filter_size 3*3, pool_size 2*2, relu, padding='same'.
2	Conv2D, filter 128, Filter_size 3*3, pool_size 2*2, relu, padding='same'.
3	Conv2D, filter 256, Filter_size 3*3, pool_size 2*2, relu, padding='same'.
4	Conv2D, filter 512, Filter_size 3*3, pool_size 2*2, relu, padding='same'.
5	Conv2D, filter 512, Filter_size 3*3, relu, padding='same'.
6	Conv2D, filter 512, Filter_size 3*3, pool_size 2*2, relu, padding='same'.

missing and in 6 cases the precontracts sequence is missing. The missing sequences are replaced with FLAIR sequences so that all images are 3-channel. The masks are single-channel binary images. They segment the FLAIR abnormalities present in the FLAIR sequence (available for all cases). The dataset is organized into 110 volumes named by case identifier and contains information about the institution of origin. Each folder contains MR images with the following naming convention: 'TCGA_<organization code>_<patient ID>_<slide number>.tif'. The data was split using the train-test split approach, with 2000 images utilized for model training, 700 for model validation, and 390 for model testing.

3   Results of Experiments

For the experiment, a Colab GPU for 12 GB of RAM was used. Python was used to create the aforementioned GAN U-Net architecture because to its widespread use and as well as simple-to-use built-in libraries. The Keras and Tensorflow libraries are used to implement the architecture. The Adam approach was used to improve the network, using a 0.0003 learning rate and at the rate of decay of 0.5, because it performed better in our tests. Training took up to 5 h for the model we provided in relation to our dataset, it was trained using 40 epochs and three brain MR images per batch and a 512 * 512 dimension We used a dice score coefficient (DSC) to evaluate our segmentation network, which is a frequently used evaluation measure for image segmentation. The following equation determines the dice score:

$$DSC(X, Y) = \frac{2 \times |X \bigcap Y|}{|X| + |Y|}. \tag{2}$$

where X indicates the predicted set of pixels and Y indicates the actual set of pixels.

We were successful in obtaining a dice score of 0.97 with our model, which is more efficient and performant than other techniques. Despite the fact that the tests were conducted on different data sets, our model performed admirably. Our model's training and validation dice score coefficients are shown in Fig. 4.

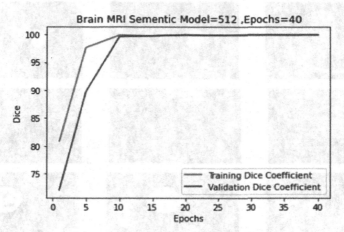

Fig. 4. Dice score coefficient of our model.

The results of the training and validate our model has given the best performance. Another observation is from epoch 16 our model is convergent and a good fit.

Table 3 displays the results of our comparison of different topologies in our dataset. The suggested model has also surpassed existing the most up-to-date methods in terms of segmentation performance for our data.

Table 3. Comparison of approaches

Model name	Dice coefficient
CNN [19]	0.840
FCNN [20]	0.865
Baris Kayalibay [18]	0.90
The proposed architecture	0.97

To test our model, we tested it on a dataset of 390 images on which it had never been trained before, and we obtained a cube value of 0.97 with a loss coefficient of 0.030, which is considered more efficient and powerful. Figure 5 shows our model's prediction results.

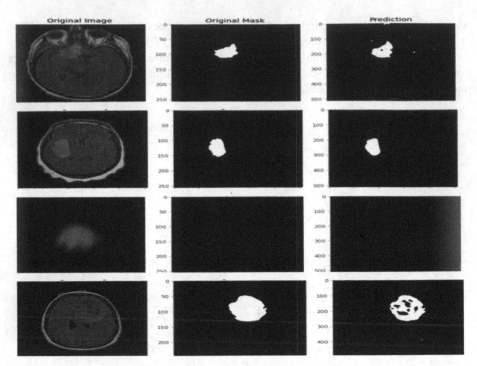

Fig. 5. Results of prediction in our model.

4   Conclusion

In our study, we present implements a segmentation the of brain cancer tumor images by GAN-U-Net architecture. Our method has achieved a more efficient and powerful dice result than the comparable methods. The suggested generative adversarial network has the drawback of requiring a lot of computing resources as well as a significant amount of time for training. Moreover, in future work, we will develop another GAN model to improve the stability of segmentation in datasets and add different noises level to prove robustness our method with different image sizes and Compare study performance and results.

References

1. Schmidhuber, J.: Deep learning in neural networks: an overview. Neural Netw. (2014). https://doi.org/10.1016/j.neunet.2014.09.003
2. Long, J., Shelhamer, E., Darrell, T.: Fully convolutional networks for semantic segmentation. In: Proceedings of the IEEE Conference on Computer Vision and Pattern Recognition, pp. 3431–3440 (2015)
3. Ronneberger, O., Fischer, P., Brox, T.: U-net: convolutional networks for biomedical image segmentation. In: Navab, N., Hornegger, J., Wells, W.M., Frangi, A.F. (eds.) MICCAI 2015. LNCS, vol. 9351, pp. 234–241. Springer, Cham (2015). https://doi.org/10.1007/978-3-319-24574-4_28

4. Ronneberger, O., Fischer, P., Brox, T.: U-Net: convolutional networks for biomedical image segmentation. In: Navab, N., Hornegger, J., Wells, William M., Frangi, A.F. (eds.) MICCAI 2015. LNCS, vol. 9351, pp. 234–241. Springer, Cham (2015). https://doi.org/10.1007/978-3-319-24574-4_28

5. Anantharaman, R., Velazquez, W., Lee, Y.: Utilizing mask R-CNN for detection and segmentation of oral diseases. In: 2018 IEEE International Conference on Bioinformatics and Biomedicine (BIBM), pp. 2197–2204 (2018). https://doi.org/10.1109/BIBM.2018.8621112

6. Holland, E.C.: Progenitor cells and glioma formation. Curr. Opin. Neurol. 14(6), 683–688 (2001)

7. Xue, Y., Xu, T., Zhang, H., et al.: SegAN: Adversarial network with multi-scale L1 loss for medical image segmentation. Neuroinform 16, 383–392 (2018)

8. Dai, W., Dong, N., Wang, Z., Liang, X., Zhang, H., Xing, E.P.: SCAN: structure correcting adversarial network for organ segmentation in chest X-rays. In: Stoyanov, D., et al. (eds.) DLMIA/ML-CDS -2018. LNCS, vol. 11045, pp. 263–273. Springer, Cham (2018). https://doi.org/10.1007/978-3-030-00889-5_30

9. Russakovsky, O., et al.: Imagenet large scale visual recognition challenge. Int. J. Comput. Vision 115(3), 211–252 (2015)

10. Yang, D., et al.: Automatic liver segmentation using an adversarial image-to-image network. In: Descoteaux, M., Maier-Hein, L., Franz, A., Jannin, P., Collins, D.L., Duchesne, S. (eds.) MICCAI 2017. LNCS, vol. 10435, pp. 507–515. Springer, Cham (2017). https://doi.org/10.1007/978-3-319-66179-7_58

11. Munawar, F., Azmat, S., Iqbal, T., Grönlund, C., Ali, H.: Segmentation of Lungs in Chest X-Ray Image Using Generative Adversarial Networks (2020)

12. Shiraishi, J., et al.: Development of a digital image database for chest radiographs with and without a lung nodule receiver operating characteristic analysis of Radiologists' detection of pulmonary nodules. Amer. J. Roentgenol. 174(1), 71–74 (2000)

13. Chen, H., Qin, Z., Ding, Y., Lan, T.: Brain tumor segmentation with generative adversarial nets. In: 2nd International Conference on Artificial Intelligence and Big Data (ICAIBD), pp. 301–305 (2019). https://doi.org/10.1109/ICAIBD.2019.8836968

14. Goodfellow, J., et al.: Generative Adversarial Networks. arXiv:1406.2661 [stat.ML] (2014)

15. Goodfellow, P., Mirza, M., Xu, B., Warde-Farley D., Ozair, S., Bengio, Y.: Generative adversarial nets. In: Proceedings of Advances in Neural Information Processing Systems, pp. 2672–2680 (2014)

16. https://wiki.cancerimagingarchive.net/display/Public/TCGA-LGG

17. Krizhevsky, A., Sutskever, I., Hinton, G.: ImageNet classification with deep convolutional neural networks. In: Neural Information Processing Systems, vol. 25 (2012). https://doi.org/10.1145/3065386

18. Myronenko, A.: 3D MRI Brain Tumor Segmentation Using Autoencoder Regularization. Brainlesion: Glioma, Multiple Sclerosis, Stroke and Traumatic Brain Injuries, pp. 311–320 (2018)

19. Pereira, S., Pinto, A., Alves, V., Silva, C.A.: Brain tumor segmentation using convolutional neural networks in MRI images. IEEE Trans. Med. Imaging 35, 1240–1251 (2016)

20. Zhao, X., Wu, Y., Song, G., Li, Z., Zhang, Y., Fan, Y.: A deep learning model integrating FCNNs and CRFs for brain tumor segmentation. Med. Image Anal. 43, 98–111 (2018)

21. El Mansouri, O., El Mourabit, Y., El Habouz, Y.: System segmentation of Lungs in images chest X-ray using the generative adversarial network. In: ITM Web of Conferences, vol. 43, p. 01020 (2022)

Hospital Room Management for Covid-19 Patients Using Petri Nets

Ayoub El Bazzazi[✉], Bouchra El Akraoui, and Abdelhadi Larach

TIAD Laboratory, Faculty of Sciences and Technics, Sultan Moulay Slimane University,
Beni Mellal, Morocco
ayb.elbazzazi@gmail.com

Abstract. This paper treat the design of the sequence organization and then the optimization of a discrete event system (DES) modelled by Temporal Petri Net (T-PN) comprising a set of specifications corresponding to time intervals to activate or access another event. A Petri net is a well-known model that describes distributed systems. It is commonly used to describe various aspects of distributed systems, such as choice and synchronization.

This paper focuses on the organizing problems in the hospitalization domain during the Covid-19 pandemic. We advocate the use of a real time approach based on TemporalPN and mathematical modeling to help drive the healthcare system in the face of occurrence of this type of giving many patients currently, which requires rethinking the predictive decision. The proposed solution permits to optimize the time to find all empty rooms using PN Temporal and the Dijkstra approach.

Keywords: Covid-19 · Reachability graph RG · Discrete event system (DES) · Timed Petri Net · Petri Net · Dijkstra

1 Introduction

The world is facing an unprecedented threat. Soon, the Covid-19 pandemic spread around the world. Because of this epidemic, suffering has spread, billions of lives have been disrupted, and the global economy is threatened [19].

The Covid-19 virus is a threat to all humanity and therefore all humanity must act to eradicate it. Efforts by individual countries to address it will not be enough [21]. Even wealthy countries with strong health systems are being challenged. To win the Covid_19 must make a very strong hospital system well organized [20].

In this article, we deal with one of the problems of the functioning of the hospital system, namely the problem of planning and resource allocation. This is a problem of great importance in a context of limited resources, growth in demand for increasingly varied and demanding care, and a move towards controlling health care expenditure and improving the efficiency and productivity of the hospital sector.

We propose here to develop a management approach focused on the process expected followed by the Covid-19 patient in order to stop the virus as soon as possible and minimize his hospital stay. In particular, this approach will be illustrated in the case of

© Springer Nature Switzerland AG 2022
M. Fakir et al. (Eds.): CBI 2022, LNBIP 449, pp. 254–262, 2022.
https://doi.org/10.1007/978-3-031-06458-6_21

care processes passing through a service to realize the PCR test. Indeed, this type of process represents a good part of the hospitalization cases.

In this case, the functioning of the hospital rooms is determined by a provisional operating schedule that specifies the patients to be seen during the day as well as the different resources allocated to each procedure. However, this schedule is often not respected because of the different types of hazards that can occur. These include uncertainty in predicting operating times, unforeseen complications, and the arrival of urgent cases for same-day PCR testing. Active monitoring of suspected Covid-19 cases within the hospital. As some cases of Covid-19 develop severe or difficult breathing pneumonia that leads to a critical condition, cases are being looked for in which pneumonia does not improve as expected after treatment with antibiotics. A doctor who receives alerts of potential cases must respond quickly. If a doctor suspects Covid-19, a diagnostic test is given to anyone who has visited a suspected hospital. Then the patient is transferred with close guard to the isolation room before the final test results appear and after that, of course, if he is a virus carrier [22]. In this paper, we are particularly interested in the study of how to take into account the emergency in the operating schedule. We advocate the use of a real-time approach based on T-temporal PN and mathematical modelling to help steer the care system in the face of the occurrence of this type of hazard, which requires a rethink of the forecasting decision. At some point, scheduling problems were also other approaches that are similar to the A-star algorithm include Petri nets and various approaches that are inspired by the A-star algorithm. One of the advantages of these is that they can detect properties such as constraints and conflicts [21]. The rest of the paper is organized as follows: In Sect. 2, we define the petri network graph algorithm, the reachability graph and the subject to use the Dijkstra algorithm. In Sect. 3, we model how covid_19 patients can be managed using a petri network and other algorithms. Finally, we conclude, and we present our perspectives in future works.

2 Petri Network (Net) and Reachability Graph

2.1 Petri Network

An autonomous PN is a bipartite graph whose nodes are places and transitions connected by arcs. The set of places is finite and non-zero. Similarly, the set of transitions is finite and non-zero. The arcs are oriented. Each arc connects a place to a transition or a transition to a place. The state of a PN is defined by the vector of place markings, denoted M. This vector indicates the number of markings (or tokens) contained in each place.

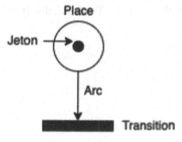

Fig. 1. Graph PN

The marks represent physical entities, for example a person, a car, a customer. An autonomous PN is formally defined as a quintuplet R = {P, T, Precedent, Post, M0} where P = {P1,..., Pn} is the places, T = {T1, ... , Tm} is the transitions, Precedent is the forward amount application, such that:

Precedent : P × T → N, (Pi, Tj) → Precedent (Pi, Tj) = the weight of the arc connecting place Pi to transition Tj, Post is the backward incidence application, such that: Post : P × T → N, (Pi, Tj) →Post (Pi, Tj) = the weight of the arc connecting transition Tj to place Pi, M0 is the initial marking.

The T-temporal PN Model

The PN T-temporal model was introduced in [19] and [20] for the modelling and analysis of communication systems. This tool is derived from the autonomous PN model (called underlying autonomous PN) by associating a time interval [aj, bj] to each transition Tj. The notions of validated transition and passable transition are no longer equivalent in a T- temporal PN model. A transition Tj is validated by the marking of the underlying autonomous PN. On the other hand, it can be crossed only when a duration included in the time interval associated with it has elapsed since the moment of its validation. Where: P = {P1 ,... Pn} is the places, T = {T1 ,... ,Tm} is the transitions, Precedent is the forward incidence application, Post is the backward incidence application, M0 is the initial marking, Is : T → Q × (Q ∪ ∞) is a function that associates a time interval with each transition. Q represents the set of positive rational numbers. Is (Ti) = [ai, bi], ai ≤ bi, 0 ≤ ai < ∞ and 0 ≤ bi ≤ ∞. The interval [ai, bi] is called the static crossing interval. It models the time constraint imposed for the crossing of the transition Ti. The time origin considered for the definition of this interval is the validation instant of the transition Ti. ai is the earliest crossing instant, biis the latest crossing time.

2.2 Reachability Graph

Accessibility graph consists of VERTEX set V and EDGE set E is designed for G = (V, E), the definition relationship is accessible to G is the closure of E, any group of all pairs (S, T) of the fifth peaks there is a series From v = s, peaks, v1, v2, ..., vk = t, so that the sixth edge, VI-1 will be in E for all <K [12].

From the marking of the PN, we can study the dynamic behavior of the modelled system. Indeed, the behavior of an autonomous PN is determined by the evolution of the marking generated by the crossing of transitions.

The evolution of an autonomous PN depends only on the marking. Therefore, an autonomous PN allows to represent only the logical sequencing of the events that intervene in the functioning of the system. To detail the transition sequence we use the Reachability graph.

Algorithm (Construction of the reachability graph of a PN)

$$GA(\,M_0(n,p),n_T,n_P,W(n_P,n_T))$$

| i=1, ... , $n_T\,n_T$

| | j=1, ..., $n_P n_P$

| | $V(j)= M_0(j)+ W(j,i)\,M_0(j)+ W(j,i)$

| | **If** $V(j) \le 0$ **SO**

| | | **Transition NO Firing**

| | | **Break**

| **IF (Firing) SO**

| | $GA(V,n_T,n_P,W,n_T,n_P,W)$

Consider the construction of the PN in Fig. 1 by using the above algorithm. For instance, if the initial is M0(1,0,0,0,0,0), there is only one possible firing transition t1, the marking obtained is M1(0,1,1,1,0,0). The accessibility graph generated by the proposed algorithm is unique. The returned graphical accessibility architecture contains everything I would tag reachable markings (Fig. 2).

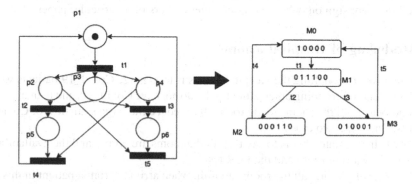

Fig. 2. Extract the accessibility graph from the Petri Net

The objective of the reachability graph is to see all the states of the module and starts from the initial mark until the final mark uses the crossable transitions. We can find a short time in our module using Dijkstra.

2.3 Dijkstra the Shortest Path Problem

Dijkstra's algorithm allows us to solve an algorithmic problem: the shortest path problem. This problem has several variants. The simplest one is the following: given an undirected graph, whose edges have weights, and two vertices of this graph, find a path between the two vertices of the graph, of minimum weight. Dijkstra's algorithm allows to solve a more general problem: the graph can be directed, and we can designate a single vertex, and ask to have the list of shortest paths for all the other nodes of the graph [16]. The algorithm uses the following auxiliary functions.

2.4 The Choice of the Petri Net Tool

It is a graphical and mathematical modelling and analysis tool widely used for the study of Discrete Event Systems (DES). However, the autonomous PN model only offers qualitative analysis possibilities. Some extensions to this model have been proposed to explicitly take into account the passage of time and allow for an analysis and quantitative evaluation of the system studied [12]. The T-temporal PN model [13] was derived from the autonomous PN model by associating to each transition a temporal constraint on the date of its crossing. Its evolution is thus determined by a single type of event: the crossing of transitions. **The choice of the T-temporal PN Tool is Therefore Justified:**

– On the one hand, by its ability to represent the main mechanisms in complex Discrete Event Systems such as care production systems (parallelism, synchronization, resource sharing, dynamic aspect and randomness). Indeed, we recognize here the characteristics of the operating room: sharing of resources during the realization of the interventions, need for synchronization between the resources, dynamic aspects imposed by the operating times and hazards occurring on these times.
– On the other hand, this tool allows to model most of the temporal constraints intervening in the operation of the operating room and offers possibilities of modelling conflicts, may sign on behalf of all the other authors of a particular paper.

3 Modelling of Hospital Rooms

The objective of modelling the Hospital rooms is to manage Covid_19 patients well to avoid problems of grouping other patients who do not have Covid_19, and to manager the time and benefit all the rooms in the room all to reduce the large statistics for Covid_19 patients. There are two cases:

Case1: Initial state, that is to say that all the rooms are empty and the availability of places in the waiting room and the PCR test.

Case2: finally State: all the rooms are full, when arriving from a patient in this case must do the confinement at home or to go from another hospital center (Fig. 3).

The different operating times are represented by intervals of the type $[\alpha, \beta]$ where α and β are whole numbers (the different operating times are expressed in days). These intervals encompass the durations of the various operations of the operating program. For example, when the patient taking the PCR test will find the result after at least 1/4 day or 2 days depending on the quality of the test. If the result of the test has not yet been found, these durations (duration of preparation of the room, installation and induction of the patient are the same as those considered during the construction of the provisional operating program. For an ongoing intervention embodiment, some of these durations may be re-estimated depending on the evolution of the patient's condition, in particular in the case of a complication.

After the arrival of the N Covid_19 patients and the K provision of the PCR test and the provision of a J placed in the waiting rooms to check whether the result is positive or negative according to the test, T2 and T3 transitions are two transitions in conflict. The choice between these two transitions depends on the quality of the test but when the patient is in an emergency, priority is thus given to the T2 transition. If the result is

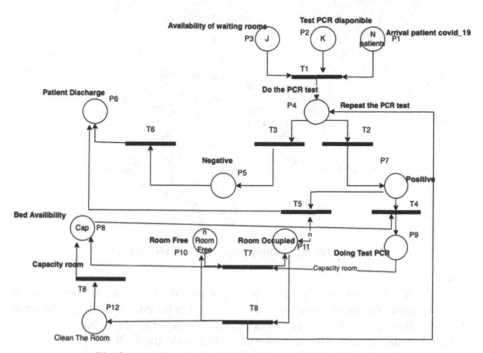

Fig. 3. Modeling the Covid_19 patients using the Petri Net.

negative, the patient can be discharged from the hospital, and if the result is positive, the patient must commit to spending at least 7 days in the hospital room if the room capacity is full. should be transferred to another hospital or confined to their home.

We need to manage all rooms to see if there are free rooms or not and can manage any room. In this case we use a reachability graph to clearly visualize all states of our model. We have two important states: State initial and final, the final state depends on the initial case, i.e. must know the stocks of the PCR tests and the availability of the waiting room and the number of Covid_19 patients who have arrived.

Example 1: Consider PN_T 1 in Fig. 4 as an example of PN_T which can find the reachability graph. All transitions are T = {t1, t2, t3, t4, t5, t6, t7, t8, t9}. The initial marking is Mi = (1 1 1 0 0 0 0 1 0 1 0 0) P = p1, p2, p3, p8, p10, and for the final marking is Mf = (0 0 0 0 0 0 0 0 0 1 0) = P11, that is to say that all the rooms are occupied in this case we only have one room. We use a reachability graph algorithm to see all possible states.

So for the first example, which is easy in the final state, we consider the arrival of a one patient Covid_19 and we have just one room and their capacity is one bed. From the reachability graph you can access any state of the module, that is to say any time you can see whether the rooms are empty or not. We can make the statistics on the beds that have remained empty or even approximate the numbers of the possible PCR test. Are they enough to manage all Covid_19 patients? What comes after all that to avoid the problems of breaking down?

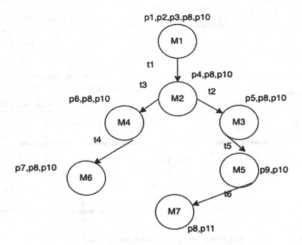

Fig. 4. The Reachability graph for all rooms is empty.

Example 2: We consider when a Covid_19 patient arrives when all the rooms are full. In this example, the patient must be referred to another hospital to do the confinement or do it at their home. When free rooms or beds in case either the rooms have not yet saturated or the patients who have already existed redo the PCR test is the result is negative in this case of course the patient must leave the room, we have need to declare that the room is empty and must manage them for other patients. In general we need to find the shortest way for the chambers which are empty (Fig. 5).

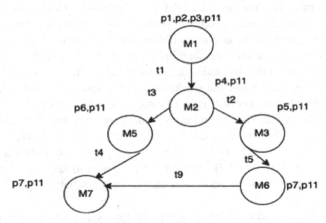

Fig. 5. All rooms of the hospital are in patients Covid_19.

The Propositions to Use on the Reachability Graph

From the sequence of transitions for the patient management Covid_19 to the hospital wards. Is to find the shortest path from the initial state when the Covid-19 patients arrive at the waiting room who want to do the test until the final state when all the rooms are occupied.

For the first example, we use the **Dijkstra** algorithm to find the nearest empty bed to a room.

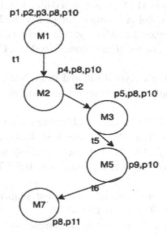

Fig. 6. Find the nearest empty bed in a room

The Fig. 6 shows the shortest path to find the empty beds in a room as fast as possible using a Dijkstra between two states and initial state when a patient Covid_19 arrives and has to do the containment of the room.

4 Conclusion

This paper has proposed a systematic design of management sequences for time delay DES; specifically, the organizing problems in the hospitalization domain during the Covid 19 pandemic. Therefore, the proposed solution permits to optimize the time to find all empty rooms using PN Temporal and the Dijkstra approach. Our future work includes assessing the risk of hitting deadlocks in order to integrate it into the control strategy and further combining Petri nets with Markov decision processes.

References

1. Wang, Q., Wang, Z.: Hybrid heuristic search based on Petri net for MS scheduling. Energy Procedia **17**, 506–512 (2012)
2. Chang, Y., Lin, C., Tsain, J., Hing, C.T.: Infection control measure of a Taiwanes hospital to confront the Covid_19 pandamic. Med. Sci. **19** (2020)
3. Lei, H., Xing, K., Han, L., Xiong, F., Ge, Z.: Deadlock-free scheduling for flexible manufacturing systems using Petri nets and heuristic search. Comput. Ind. Eng. **72**, 297–305 (2014)
4. Xiong, H.H., Zhou, M.: Scheduling of semiconductor test facility via Petri nets and hybrid heuristic search. IEEE Trans. Semicond. Manuf. **11**(3), 384–393 (1998)

5. Jeng, M.D., Chen, S.C.: Heuristic search approach using approximate solutions to Petri net state equations for scheduling flexible manufacturing systems. Int. J. Flexible Manuf. Syst. **10**(2), 139–162 (1998)
6. Zhang, W., Freiheit, T., Yang, H.: Dynamic scheduling in a flexible assembly system based on a timed Petri nets model. Robot. Comput. Integr. Manuf. **21**(6), 550–558 (2005)
7. Mejía, G., Caballero-Villalobos, J.P., Montoya, C.: Petri nets and deadlock-free scheduling of open shop manufacturing systems. IEEE Trans. Syst. Man, Cybern. Syst. **48**(6), 1017–1028 (2018)
8. Mejíaand, G., Niño, K.: Anew Hybrid Filtered Beam Search Algorithm for deadlock-free scheduling of flexible manufacturing systems using Petri nets. Comput. Ind. Eng. **108**, 165–176 (2017)
9. Pan, L., Ding, Z.J., Zhou, M.C.: A configurable state class method for temporal analysis of time Petri nets. IEEE Trans. Syst. Man Cybern. Syst. **44**(4), 482–493 (2014)
10. Zhu, Q., Zhou, M., Qiao, Y., Wu, N.: Petri net modelling and scheduling of a close-down process for time-constrained single-arm cluster tools. IEEE Trans. Syst. Man Cybern. Syst. **48**(3), 389–400 (2018)
11. Tarek, A., Lopez-Benitez, N.: Optimal legal firing sequence of Petri nets using linear programming. Optim. Eng. **5**(1), 25–43 (2004)
12. Hu, H., Su, R., Zhou, M., Liu, Y.: Polynomially complex synthesis of distributed supervisors for large-scale AMSs using Petri Nets. IEEE Trans. Control Syst. Tech. **24**(5), 1–13 (2015)
13. Lefebvre, D., Leclercq, E.: Control design for trajectory tracking with untimed Petri nets. IEEE Trans. Autom. Control **60**(7), 1921–1926 (2015)
14. Lefebvre, D.: Approaching minimal time control sequences for timed Petri nets. IEEE Trans. Autom. Sci. Eng. **13**(2), 1215–1221 (2016)
15. de Albuquerque, G.A., Maciel, P., Lima, R.M.F., Magnani, F.: Strategic and tactical evaluation of conflicting environment and business goals in green supply chains. IEEE Trans. Syst. Man Cybern. Syst. **43**(5), 1013–1027 (2013)
16. Lefebvre, D., Daoui, C.: Control design for untimed Petri nets using Markov decision processes. J. Oper. Res. Decis. **27**(4), 28–43 (2017)
17. Berthomieu, B., Menasche, M.: An enumerative approach for analysing time Petri nets. In: Proceedings of IFIP Congress, pp. 41–46 (1983)
18. Berthomieu, B., Vernadat, F.: State class constructions for branching analysis of time Petri Nets. In: Garavel, H., Hatcliff, J. (eds.) Tools and Algorithms for the Construction and Analysis of Systems. TACAS 2003. Lecture Notes in Computer Science, vol. 2619. Springer, Berlin (2003). https://doi.org/10.1007/3-540-36577-X_33
19. Gardey, G., Roux, O.H., Roux, O.F.: Using Zone graph method for computing the state space of a time Petri net. In Formal Modelling and Analysis of Timed Systems (2000)
20. Wu et Wu, S.D., Storer, R.H., Chang, P.C.: One machine rescheduling heuristics with efficiency and stability as criteria. Comput. Oper. Res. **20**, 1–14 (1993)
21. Lube, J., Awel, D., Mogga, J.: 17 merch 2021 COVID-19 case management strategies: what are the options for Africa. Infectious Diseases of poverty. N 30(2021). 17 Mar 2021

Dimensionality Reduction of MI-EEG Data via Convolutional Autoencoders with a Low Size Dataset

Mouad Riyad[✉], Mohammed Khalil, and Abdellah Adib

LMCSA Laboratory, Faculty of Sciences and Technology,
Hassan II University of Casablanca, Mohammedia, Morocco
riyadmouad1@gmail.com, mohammed.khalil@univh2c.ma, adib@fstm.ac.ma

Abstract. Electroencephalography has a poor spatial resolution, as experimental setups demand many electrodes around the motor cortex to reach the best results. Yet, it increases the data to be stored or transmitted in real-time for later uses. Thus, researchers have suggested autoencoders (AE) that transmit the compressed latent variable instead of the data itself. In this paper, we propose an AE and a Supervised Autoencoder (SupAE) designed for mobile applications treating Motor Imagery (MI). The introduced Encoder and Decoder derive from the previously published AMSI-EEGNet, a fast-to-train and lightweight architecture. The results found that the proposed methods perform better than baselines, especially for a high compression ratio (CR). Also, SupAE is a better option when the transmitted data needs classification. Further, we studied the evolution of the AE training and found that it learns similar features to previous studies.

Keywords: Data compression · Brain-computer interface · Electroencephalography · Convolutional neural network · Autoencoder

1 Introduction

Medical data is omniscient nowadays, as many institutions collect medical exams results for archiving or for Machine Learning systems [12]. For this reason, they keep a large amount of data stored for later use. Electroencephalography (EEG) is a non-invasive imaging technique of the brain [26]. It permits the diagnosis of several neural disorders or enables non-medical applications for entertainment [8,17]. It requires sensing the neural activity through an array of sensors placed on the head's scalp. Motor Imagery (MI) are waves produced voluntarily by a person by imagining performing a movement with a limb [4]. It produce μ and β waves located respectively in the segments 8–13 Hz and 13–30 Hz according to many studies [22].

For biomedical data, there is a necessity to gather most data as possible to improve machine learning models, given that they can save many lives [21].

M. Fakir et al. (Eds.): CBI 2022, LNBIP 449, pp. 263–278, 2022.
https://doi.org/10.1007/978-3-031-06458-6_22

The most interesting example is that EEG is a front-line diagnostic tool for brain disease due to its low cost, ease of setup, and non-invasive sensors. Sometimes, the number of electrodes can reach 128 electrodes, that sample at a high rate in some setups. In those cases, the compression of the signals becomes an obligation to save storage space or speed up data transfer. The easiest solution is dropping unnecessary electrodes according to the case or lowering the sampling rate whenever is possible. Also, it is possible to use standard compression algorithms on binary files, such as Lempel-Ziv or Lempel-Ziv-Welch algorithms. Others have taken advantage of the nature of data where used well-known audio compression strategies relying on transform such as Discrete Fourier Transform (DFT), Discrete Cosine Transform (DCT), Discrete Wavelet transform (DWT) [6].

Although this may be true, Deep Learning (DL) breakthrough offers new insight into several aspects of data analysis [16]. At first, it is a framework that allows the artificial Neural Network (NN) to learn and extract knowledge from the data. However, it demands more data than the traditional machine learning algorithms. Hopefully, it enables many use-cases such as data compression [2]. The basic idea is to use AutoEncoders (AE) (which is a special neural network) that encodes any data into a condensed representation and that it can decode it into its original shape [10]. Several variants were proposed, including Denoising-Autoencoder (DAE), Variational-Autoencoder (VAE), and Supervised-Autoencoder (SupAE) [14]. For EEG applications, some research investigated the compression of EEG in a few experiments [3,20]. However, literature methods have little study its application on MI and more precisely on epilepsy waves [9,11,19]. Equally important, they applied most models on big datasets and were preprocessed, unlike the case of real-life implementation.

In this paper, we propose a novel autoencoder design for EEG signals compression. We introduce an Encoder-Decoder architecture with low computational power and a few parameters. Thus, we use convolutional layers inspired by MobileNet and related state-of-the-art architectures applied for EEG. Compared with the existing studies, we study the implication of compression of MI on low-sized datasets with a light preprocessing procedure. Further, we propose the SupAE to reduce the constraint of decompressing and classify the compressed data directly as in [7].

We propose the following organization for this paper: Sect. 2 provides the introduction of the concept of autoencoder, along with the new encoder and decoder architectures. Section 3 presents the experimental setup and results. In Sect. 4, we analyze the obtained results. Section 5 summarizes the finding of this paper.

2   Methods

2.1   Autoencoders

AE are neural networks used for feature learning and dimensionality reduction in an unsupervised process [10]. Their main goal is to learn a compressed representation of the data that keeps the essential features, which we can decompress afterward to reconstruct the same input. The AE has two parts: The encoder that maps an input x into a latent variable z and the decoder that reconstructs the input into x'. The architecture creates a bottleneck scheme by making the dimensions of the latent representation z smaller than the input. It is described by the formulas 1 and 2:

$$z = q_\theta(x) \tag{1}$$

$$x' = p_\phi(z) \tag{2}$$

where q represents the mapping into the latent variable and θ the learned weight of the encoder, when p represents the mapping into the reconstruction and ϕ the decoder parameters. The parameters θ and ϕ are learned, by minimizing a loss function $L(.)$, through Stochastic Gradient Descent as used for most NNs. There are several types of autoencoders, and each one of them aims to learn the optimal weight through different strategies, as shown in Fig. 1:

Vanilla Autoencoder. The most basic AE seeks to minimize the reconstruction error between x and x' by using the Mean Square Error (MSE) as the Eq. 3 shows.

$$L(x, x') = MSE(x, x') = \| x - x' \|^2 \tag{3}$$

This loss function ensures that the autoencoder is sensitive to the input and ensures that it learns its features. We can add other terms to this loss function to ensure other objectives depending on the application.

If we corrupt the input with noise or use dropout in the encoder, the AE will behave like a DAE. However, the purpose of the latter one is to restore corrupt input, which is not the scope of this paper. But, we will use dropout afterward in this paper at a small rate for better generalization due to the small dataset. It must be remembered that other autoencoders may not be relevant for this paper, as we seek efficiency in memory usage and training time.

Supervised Autoencoder. We can use autoencoders for classification by adding a term that secures the discrimination between classes and improves disentangling the latent variable [15]. It is important to note that it allows, by its design, to classify data directly from the transferred latent variable without need to decode it [28]. In fact, there is no need to decompress the data. We must therefore use the Cross Entropy (CE) loss function which is used for classification problems, as shown in Eq. 4.

$$CE(y, y') = -\sum_{i=1}^{C} y_i' log(y) \tag{4}$$

where y represents the true class of the sample and y' the estimated one. However, we add a regularization variable α to control the trade-off between the reconstruction and the classification.

$$L(x, x', y, y') = \alpha \times MSE(x, x') + CE(y, y') \tag{5}$$

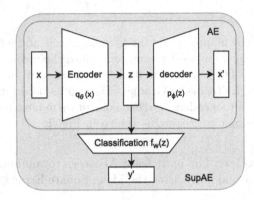

Fig. 1. Architecture of the Autoencoders structure

2.2 Encoder-Decoder

The main contribution of this paper is the encoder-decoder design, considering complications cited in the literature. EEG signals are high dimensional compared with other biosignals because of their multichannel configuration. Also, we aim to design a light architecture that permits fast encoding and decoding. We use a ConvNet configuration with no dense layer to reduce encoder-decoder parameter size and avoid overfitting. The architecture will be based on a multiscale ConvNet as in [1,23] and represented in Fig. 3. Thus we suggest the following configuration:

For the encoder, the input gets a multichannel EEG matrix-shaped as a matrix of size (C, T) where C represents the number of channels and T the timestamp. The matrix will go through two paths: The main one will keep the original data, and the other path produces downsampled versions of the signals at $fs/2$, $fs/4$, and $fs/8$. The building blocks of our approach are shown in Fig. 2. The units called Spatio-Temporal Block (STB), displayed in Fig. 2a, are fed with the signals (original and downsampled versions) [13]. It has a convolutional layer with F kernels with a kernel size of $(1, 3)$. Then we use spatial filtering through a depthwise convolutional layer with a depth parameter of D and a kernel size of $(1, C)$. After each convolutional layer, we use batch norm layers and activation after the second one only. As regularization measures, we adopt dropout layers at the end of the block and a weight norm constraint for the second layer. Afterward, each layer output is added progressively to the Aggregative Block (AB), as in Fig. 2b, until we assemble the latent variable. The latent variable

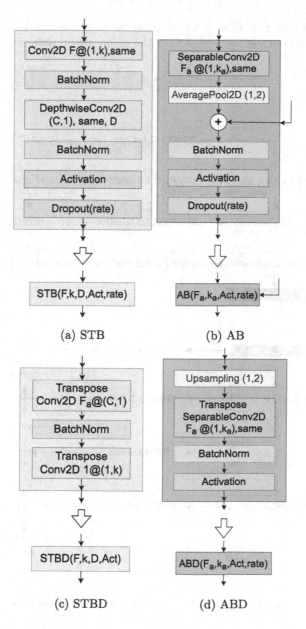

Fig. 2. The basic blocks used to built our Autoencoder

size will be of size $F \times D \times \frac{T}{8}$ which must be kept the size below $C \times T$. The reduction of the data is measured with the compression defined as in Eq. 6

$$CR = \frac{OriginalSize}{CompressedSize} = \frac{C \times T}{F \times D \times \frac{T}{8}} = \frac{8 \times C}{F \times D} \qquad (6)$$

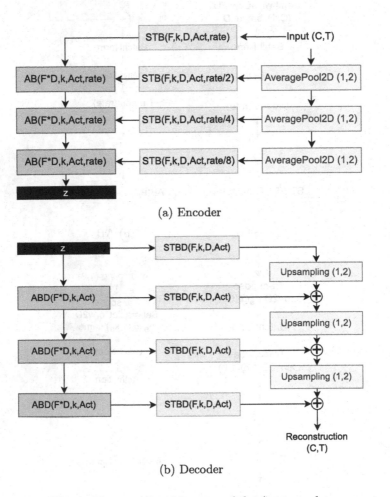

(a) Encoder

(b) Decoder

Fig. 3. The overall architecture of the Autoencoder

For the decoder, we design a symmetric structure based on the encoder. Each block of the encoder has an equivalent in the decoder and takes the same hyperparameters. The decoder does not contain any regularization, such as the dropout layer. We adopt a Transpose Separable Convolutional layer to reduce the number of parameters. This Convolution has a Transpose Depthwise Convolution followed by a Transpose Pointwise Convolution. To reverse the STB block, we proposethe Spatio-Temporal Block for Decoder (STBD). It has two Transpose

Convolutional Layer separated by a batch norm layer as displayed in Fig. 2c. Meanwhile, the Aggregative Block for the Decoder (ABD) inverts the AB effect containing an upsampling layer, Transpose Convolutional layer, BatchNorm, and activation as in Fig. 2d. The decoder takes the latent variable as an input. It is fed simultaneously to an STBD and an ABD block. The STBD goal is to turn a representation into an EEG-like signal while the ABD upsample the data. We sum the STBD outputs as each one passes its one (after upsampling it) to the other one of another STBD with a higher sampling frequency. The ABD transmits its output to another ABD for further upsampling. The decoder will keep with these steps until it composes a signal with the same dimension as the original input given to the encoder.

In current work, we neglect the Dense layer as we observed overfitting during the training and an increase in the model size. As Activation, we use the Exponential Linear Unit (ELU) in the encoder, and the Leaky Rectifier Linear Unit (LeakyReLU) for the decoder [5,18]. For the SupAE, we use a classification layer using N neurons where N is equal to the number of classes of the problem.

3 Results

3.1 Dataset

In this experiment, we use the BCI Competition IV's IIa dataset. It is constituted of a four classes motor imagery paradigms with twenty two electrodes and signal sampled at 250 Hz. The classes are related with limb movement: Left Hand (LH), Right Hand (RH), Feet (Ft), and Tongue (Tg). The dataset has two subset recorded in different sessions with equivalent proportion of data: 72 samples of each tasks and 288 samples for each session.

3.2 Experimental Setup

As preprocessing, we filter the data with a bandpass filter with cutoff frequencies 0.5 and 40 Hz. Also, we normalize the data with the same algorithms proposed by [24]. We keep segments with a length of 4s. We procede with a data augmentation strategies by adding segments with an offset of –0.2, –0.1, 0.1 and 0.2. Thus we have 1440 samples for training and 288 for testing.

To train our methods, we train AE models for 120 epochs and the SupAE models for 300 epochs. We use Adam Optimizer with a learning rate of 10^{-3}. We fix the β parameter and the dropout rate to 0.05 (There is no link between the two parameters). We test a few sets of the F-D parameters to achieve distinct CR values. The sets 11-4, 10-2, and 8-2 are associated respectively with the CRs 4, 8, and 11. We coded the experiment with Python 3 and used Pytorch 1.10 for the training on a P100 GPU.

To evaluate our approaches, we design baselines to evaluate our methods based on basic and known methods. We propose to use DWT, where we decompose our signals, discard some decompositions and reconstruct the signal with the inverse transform. Also, we propose an approach based on interpolation (Inter) by downsampling the signal and upsample it back.

3.3   Evaluation

Table 1. Compression performance of the proposed methods and the baselines in terms of MSE

	CR									
	4				8				11	
	DWT	Inter	AE	SupAE	DWT	Inter	AE	SupAE	AE	SupAE
S1	**0.086**	0.153	0.113	0.111	0.213	0.430	**0.154**	0.160	**0.145**	0.213
S2	**0.209**	0.371	0.276	0.265	0.447	0.881	**0.345**	0.421	**0.309**	0.463
S3	0.145	0.257	**0.102**	0.122	0.343	0.688	**0.199**	0.213	0.256	**0.232**
S4	**0.055**	0.098	0.066	0.074	0.151	0.309	**0.094**	0.110	**0.083**	0.169
S5	**0.096**	0.172	0.102	0.138	0.231	0.459	**0.161**	0.234	**0.162**	0.220
S6	**0.062**	0.110	0.074	0.093	0.180	0.374	**0.105**	0.158	**0.138**	0.237
S7	**0.049**	0.088	0.065	0.064	0.144	0.293	**0.086**	0.091	**0.089**	0.113
S8	0.129	0.228	0.102	**0.097**	0.322	0.650	**0.145**	0.159	0.194	**0.187**
S9	0.055	0.098	0.055	**0.051**	0.178	0.375	0.083	**0.080**	**0.080**	0.119
Avg	**0.099**	0.175	0.106	0.113	0.246	0.495	**0.152**	0.181	**0.162**	0.217

Table 1 presents the average MSE achieved by each technique, along with its CR. In the cases of the CR of 4, the results show that the DWT performs better than the alternative methods, but with a slow margin for the proposed methods. AE and SupAE offer great results for three subjects, with a slight advantage to AE. Besides, Inter received the worst results in this scheme. As we increase the CR to 8, we note an enhancement in the performance of deep methods as AE exceeds the other methods. The SupAE followed with the second-best value, even though the difference in the average MSE is larger this time. The DWT got the third-best value that is close to SupAE one while Inter operates badly. If we increase the CR to 11 for the proposed methods, their MSEs are still lower than DWT at 8 of CR. Also, the MSE increases slowly across CR for AE in contrast to SupAE, DWT, and Inter.

Table 2. Classification results after compressing the signals

		Methods			
		DWT	Inter	AE	SupAE
CR	4	76.5	73.6	55.3	74.8
	8	56.9	57.5	43.9	74.3
	11	-	-	42.3	72.0

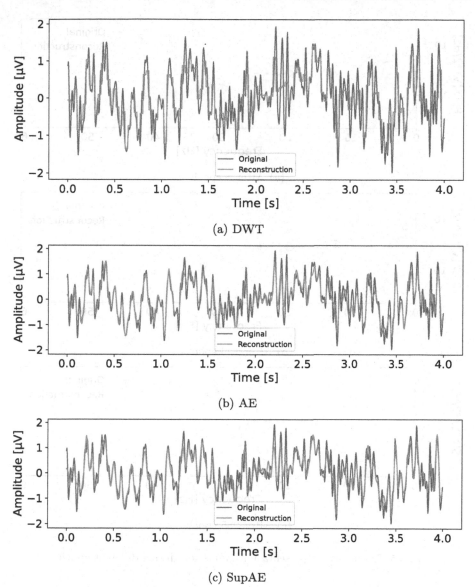

(a) DWT

(b) AE

(c) SupAE

Fig. 4. Reconstruction evaluation of different methods with CR = 8

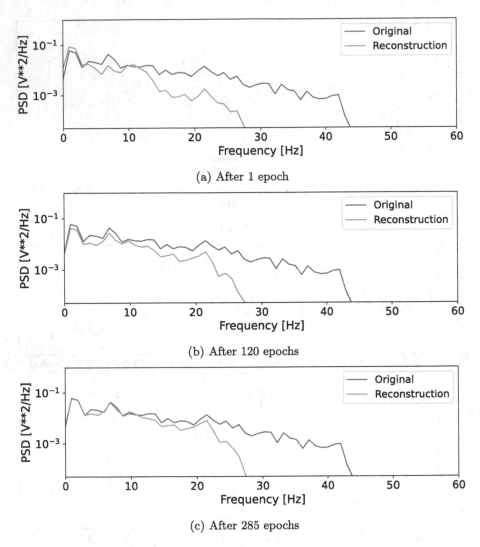

(a) After 1 epoch

(b) After 120 epochs

(c) After 285 epochs

Fig. 5. Frequential representation evaluation during different epochs

Table 2 illustrates the results of the classification after compressing the signals. In this experiment, we compress then decompress the data for all methods, we perform the classification on AMSI-EEGNet [23], except for SupAE (see Sect. 2.1). With a CR of 4, DWT achieves the highest accuracy, followed by Inter. However, SupAE had the third best result. Meanwhile, AE scored inadequate accuracy along with Inter. When we increased the CR to 8, the performances dropped for DWT and Inter. For SupAE, it performs better than the previous ones. SupAE's performance decreased slightly compared to the other methods, even up to a CR of 11.

Further, Table 3 represents a summary of the state-of-the-art related to the compression of EEG. It indicates some comparable methods achievement in terms of Peak Signal-to-Noise-Ratio (PSNR). Our methods reached a competitive PSNR compared with other works while comparing with [27] and [25]. Moreover, we could achieve a twice higher CR.

Table 3. Relative comparaison of our methods with others state-of-the-art

Paper	Method	CR	PSNR
[27]	SPC	5.4	32
[25]	MMSPC	5.03	22.21
Proposed	AE	4	33.24
		8	31.60
		11	31.34
	SupAE	4	32.93
		8	30.94
		11	29.96

For further analysis, we take electrode C3 of subject 6 while performing a Rh task. To study the quality of the reconstruction, we plotted the original signal and its reconstruction in Fig. 4, and the same for DWT, AE, and SupAE. We observe that the decompression strongly distorted the data because it could not reconstruct the low period oscillation for DWT as Fig. 4 illustrates. The reconstruction of the AE and SupAE models, shown respectively in Figs. 4b and 4c, shows better results except for the picks, as their reconstruction has lower amplitude.

Meanwhile, Fig. 5 displays the reconstruction in the frequential domain. Figure 5a shows that the model could learn the representation of the low frequencies. However, the μ and β bands are not matching. Figure 5b demonstrates that the model learned to reproduce most of the band of interest. Contrasted with the previous one, the magnitude of the PSD amplified and is matching enough of the β band and a pick in the μ band. Figure 5c exhibits an improvement in the reconstruction of low frequencies picks and a pick at the β band. Beyond 28 Hz, the model could not rebuild the high frequencies beyond the β at all stages of the learning.

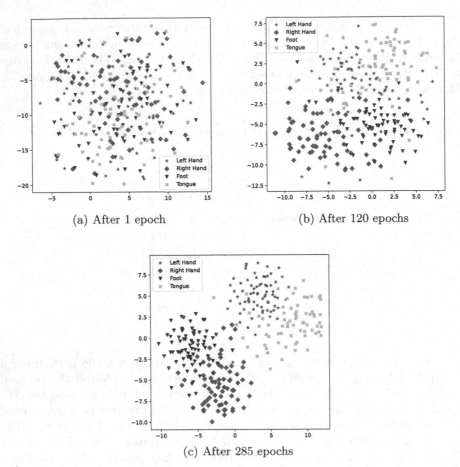

(a) After 1 epoch   (b) After 120 epochs

(c) After 285 epochs

Fig. 6. Latent variable evaluation during different epochs

Figure 6 shows the state of the latent variable at various epochs over the training process, with t-SNE for the SupAE model. We can see that, in the beginning, after the first epoch in Fig. 6a, the points are scatted in the plot, meaning that the latent variable is entangled. After 120 epochs, Fig. 6b shows good clustering of the samples. However, the boundaries are overlapping. Later, Fig. 6b displays much proper clustering between two clusters based on Lh-Rh and Ft-Tg. However, the boundaries between Lh and Rh clusters are tight. We did not consider AE models for this experiment because they similarly incorporate their latent variable in Fig. 6a during all epochs.

Table 4. Evaluation of the CR evolution after a second compression by standard algorithms

Algorithm	4		8		11	
	AE	SupAE	AE	SupAE	AE	SupAE
bz2	4.28	4.27	8.56	8.54	11.75	11.74
zlib	4.33	4.33	8.66	8.67	11.90	11.92
zstd	4.34	4.34	8.68	8.69	11.93	11.94

Furthermore, it is possible to combine our method with other compression techniques by compressing the latent variable. However, it will be difficult to use another lossy algorithm again as it may corrupt the data further. Thus, we use standard compression algorithms such as bz2, zlib, and zstd to assess the increase of CR as Table 4 depicts. We observe that the compression algorithms permit an increase in the CR for both SupAE and AE. The difference between both methods is small when we use the same compression algorithm. The zstd algorithms permitted better performance than zlib and bz2.

4 Discussion

This work aims to establish a new approach to EEG compression with better achievements. The results show that with a CR of 4, the DWT is more suitable for this task. However, with the increase of the CR, our methods took the upper hand over the alternative ones. Further, the results suggest that the MSE increases slowly compared with the traditional approaches. In addition, we evaluate the classification performance after the data being compressed. At first, SupAE, DWT and Inter achieved satisfying results while AE was unable to function properly. By increasing the CR, SupAE did lost slightly in performance compared to baseline. It is important to realize that increasing the CR implies a reduction in the complexity of the models like Eq. 6 which is a great advantage.

Further, we analyzed the learning over the training by checking the latent variable and the reconstruction. We observed that the SupAE learned to cluster the samples into two clusters: hands samples (Lh and Rh) and the others (Ft and Tg). Simultaneously, the reconstruction's frequential spectrums display correct reproduction of μ and lower β bands. Then it learns to split each class inside each cluster which implied a better reconstruction of the high β band segment. In conjuncture with results of [22], we suggest that the first disentanglement (between hands samples and the others) is due to the fact that the network learned to discriminate based on the perturbation of the μ and β bands. Then, the network learns more details to differentiate between hand movements, and between Tg and Ft which explain our observations in our previous work [23].

Given those points, we conclude that Deep methods are more suitable in implementations that require high CR especially. However, each architecture showed satisfactory results in one task better than the others. Thus further studies must investigate a trade-off between both models. VAE may be an option due to its adoption by many researchers, but it did not show relevant results to be reported. Future studies must investigate a way to make it work.

5   Conclusion

In this paper, we introduced a new convolutional autoencoder for EEG signals transmission. The main characteristic of the approach relies on its ability to perform well on high CR. Also, this approach design reduces the number of parameters of the models enabling fast inference. We also studied the learning dynamic to understand the steps followed by the model to rebuild the EEG signals. Future works may study alternative approaches based on improved variants of the proposed autoencoders without improving the combination with other techniques.

References

1. Amin, S.U., Alsulaiman, M., Muhammad, G., Mekhtiche, M.A., Shamim Hossain, M.: Deep learning for EEG motor imagery classification based on multi-layer CNNs feature fusion. Fut. Gene. Comput. Syst. **101**, 542–554 (2019). https://doi.org/10. 1016/j.future.2019.06.027
2. Ben Said, A., Mohamed, A., Elfouly, T.: Deep learning approach for EEG compression in mHealth system. In: 2017 13th International Wireless Communications and Mobile Computing Conference (IWCMC), pp. 1508–1512. IEEE, Valencia, Spain, June 2017. https://doi.org/10.1109/IWCMC.2017.7986507
3. Cao, Y., Zhang, H., Choi, Y.B., Wang, H., Xiao, S.: Hybrid deep learning model assisted data compression and classification for efficient data delivery in mobile health applications. IEEE Access **8**, 94757–94766 (2020). https://doi.org/10.1109/ ACCESS.2020.2995442
4. Clerc, M., Bougrain, L., Lotte, F. (eds.): Brain-Computer Interfaces 1: Foundations and Methods. Cognitive Science Series, ISTE; Wiley, London (2016)
5. Clevert, D.A., Unterthiner, T., Hochreiter, S.: Fast and accurate deep network learning by exponential linear units (ELUs). In: Bengio, Y., LeCun, Y. (eds.) 4th International Conference on Learning Representations, ICLR 2016, San Juan, Puerto Rico, 2–4 May 2016, Conference Track Proceedings (2016)
6. Dao, P.T., Li, X.J., Do, H.N.: Lossy compression techniques for EEG signals. In: 2015 International Conference on Advanced Technologies for Communications (ATC), pp. 154–159 (2015). https://doi.org/10.1109/ATC.2015.7388309
7. Ditthapron, A., Banluesombatkul, N., Ketrat, S., Chuangsuwanich, E., Wilaiprasitporn, T.: Universal joint feature extraction for P300 EEG classification using multi-task autoencoder. IEEE Access **7**, 68415–68428 (2019). https://doi.org/10. 1109/ACCESS.2019.2919143
8. Farwell, L., Donchin, E.: Talking off the top of your head: toward a mental prosthesis utilizing event-related brain potentials. Electroencephalogr. Clin. Neurophysiol. **70**(6), 510–523 (1988). https://doi.org/10.1016/0013-4694(88)90149-6
9. Gogna, A., Majumdar, A., Ward, R.: Semi-supervised stacked label consistent autoencoder for reconstruction and analysis of biomedical signals. IEEE Trans. Biomed. Eng. **64**(9), 2196–2205 (2017). https://doi.org/10.1109/TBME. 2016.2631620

10. Goodfellow, I., Bengio, Y., Courville, A.: Deep Learning Adaptive Computation and Machine Learning. The MIT Press, Cambridge (2016)
11. Hosseini, M.P., Soltanian-Zadeh, H., Elisevich, K., Pompili, D.: Cloud-based deep learning of Big EEG data for epileptic seizure prediction. arXiv:1702.05192 [cs, stat], February 2017
12. Kaya, M., Binli, M.K., Ozbay, E., Yanar, H., Mishchenko, Y.: A large electroencephalographic motor imagery dataset for electroencephalographic brain computer interfaces. Sci. Data 5, 180211 (2018). https://doi.org/10.1038/sdata.2018.211
13. Lawhern, V.J., et al.: EEGNet: a compact convolutional neural network for EEG-based brain-computer interfaces. J. Neural Eng. 15(5), 056013 (2018). https://doi.org/10.1088/1741-2552/aace8c
14. Le, L., Patterson, A., White, M.: Supervised autoencoders: improving generalization performance with unsupervised regularizers. In: 32nd Conference on Neural Information Processing Systems (NeurIPS 2018), Montreal, Canada, p. 11 (2018)
15. Le, L., Patterson, A., White, M.: Supervised autoencoders: improving generalization performance with unsupervised regularizers. In: Bengio, S., Wallach, H., Larochelle, H., Grauman, K., Cesa-Bianchi, N., Garnett, R. (eds.) Advances in Neural Information Processing Systems. vol. 31. Curran Associates, Inc. (2018)
16. LeCun, Y., Bengio, Y., Hinton, G.: Deep learning. Nature 521(7553), 436–444 (2015). https://doi.org/10.1038/nature14539
17. Liao, L.D., et al.: Gaming control using a wearable and wireless EEG-based brain-computer interface device with novel dry foam-based sensors. J. NeuroEng. Rehabi. 9(1), 5 (2012). https://doi.org/10.1186/1743-0003-9-5
18. Maas, A.L., Hannun, A.Y., Ng, A.Y., et al.: Rectifier nonlinearities improve neural network acoustic models. In: Proceedings of the ICML,vol. 30, p. 3. Citeseer (2013)
19. Nguyen, B., Ma, W., Tran, D.: A study of combined lossy compression and seizure detection on epileptic EEG signals. Procedia Comput. Sci. 126, 156–165 (2018). https://doi.org/10.1016/j.procs.2018.07.219
20. Nguyen, B.T.: EEG Lossy compression and its impact on EEG-based Pattern Recognition. Ph.D. thesis, University of Canberra
21. Pandey, S.K., Janghel, R.R.: Recent deep learning techniques, challenges and its applications for medical healthcare system: a review. Neural Process. Lett. 50(2), 1907–1935 (2019). https://doi.org/10.1007/s11063-018-09976-2
22. Pfurtscheller, G., Brunner, C., Schlögl, A., da Silva], F.L.: Mu rhythm (de)synchronization and EEG single-trial classification of different motor imagery tasks. NeuroImage 31(1), 153–159 (2006). https://doi.org/10.1016/j.neuroimage.2005.12.003
23. Riyad, M., Khalil, M., Adib, A.: A novel multi-scale convolutional neural network for motor imagery classification. Biomed. Signal Process. Control 68, 102747 (2021). https://doi.org/10.1016/j.bspc.2021.102747
24. Schirrmeister, R.T., et al.: Deep learning with convolutional neural networks for EEG decoding and visualization: convolutional neural networks in EEG analysis. Hum. Brain Mapp. 38(11), 5391–5420 (2017). https://doi.org/10.1002/hbm.23730
25. Sudhakar, M.S., Titus, G.: Computational mechanisms for exploiting temporal redundancies supporting multichannel EEG compression. In: Paul, S. (ed.) Application of Biomedical Engineering in Neuroscience, pp. 245–268. Springer, Singapore (2019). https://doi.org/10.1007/978-981-13-7142-4_12
26. Teplan, M., et al.: Fundamentals of EEG measurement. Measure. Sci. Rev. 2(2), 1–11 (2002)

27. Titus, G., Sudhakar, M.S.: A simple and efficient algorithm operating with linear time for MCEEG data compression. Austral. Phys. Eng. Sci. Med. **40**(3), 759–768 (2017). https://doi.org/10.1007/s13246-017-0575-x
28. Wu, D., Shi, Y., Wang, Z., Yang, J., Sawan, M.: C^2SP-Net: joint compression and classification network for epilepsy seizure prediction. arXiv:2110.13674 [cs], October 2021

Car Tracking Technique for DLES Project

Abderrahman Azi$^{(\boxtimes)}$, Abderrahim Salhi , and Mostafa Jourhmane

Information Processing and Decision Laboratory, Sultan Moulay Slimane University,
Beni-Mellal, Morocco
abderrahman.azi@gmail.com, ab.salhi@gmail.com,
m.jourhmane@usms.ma

Abstract. Object tracking problem is considered as one of interesting area of research due to the large utilization of video surveillance systems these days. In this paper, a real-time tracking method based on HSV color space is proposed for car location detection in a specific area applied on surveillance videos. This method begins with pre-processing step in which a 3D to 2D transformation on the extracted tracking area is made in order to get the bird's eye view, alongside a bilateral filter or so-called edge preserving filter is used to reduce noises. Then, the following step is extracting the car's movement information by computing Structural Similarity Index between pre-extracted patches from frames side by side with Morphological operation for more noise removing. At the end, the obtained result is sent to a post-processing step in order to detect the location of the car and save their path for further treatment. The main goal of using this approach is to supervise Car Reverse Test (CRT) for driving license exam in Morocco which is the second part of the project called Driving License Exam Supervisor (DLES). The performance of the presented tracking system is evaluated on real surveillance scenes and shown in the experiments results. Experimental result demonstrates that the proposed method can generate a path similar to which is made by the vehicle with considerable level of efficiency for all possible situations compared to the ground truth.

Keywords: Object tracking · Structural similarity index · Car detection · Surveillance system · LMA · National road safety strategy · Driving license exam supervisor

1 Introduction

The adoption of automation approach of several activities and services particularly provided by public administrations of developing countries like Morocco has become a major priority for those nations, especially with corona virus situation and his negative impact and consequences.

The National Road Safety Strategy (2017–2026) introduced by Moroccan government came up with many arrangements and missions serving a main cause: reducing the rate of road traffic deaths with 50% (1900 death) in 2026 compared with the reference of year 2017 (3776 death). Therefore, this strategy is principally based in promoting

M. Fakir et al. (Eds.): CBI 2022, LNBIP 449, pp. 279–293, 2022.
https://doi.org/10.1007/978-3-031-06458-6_23

researches in automation of several activities and services related to the same subject and also upgrading the Information System already in use by their establishments.

Maintaining and objectively testing the driving abilities of future drivers is within headlines of the Road Safety Strategy. Moreover, these actions would raise the eligibility of the Moroccan driving license and facilitates the recognition of this license by other countries.

The driving license exam as it is described in our previous paper [1] is constituted of two sections:

Theoretical test: a question/answer quiz test where the objective is to test the candidate's knowledge before driving the vehicle.

Practical test: testing the parking, driving and controlling abilities of the candidate on the field.

The Practical Test (for car driving category "B"): This test is consisted of three sections: Car Reverse Test, Car Entering to the Garage Test, Car Parking Test and Car Driving Test.

Our project which is based on the concept of video surveillance system was proposed to manage the practical test. This kind of system has many applications, which are generally based on the event detection and object tracking analysis technique. In the first part of our proposed DLES system, we have focused on the first car reverse test (CRT System). In which, a motion detection method in the line area named LMA is proposed in order to detect any changes that occur when the vehicle crosses the start line or arrives to the end line. This technique should helps to avoid a continuous video recording and processing of static scenes. Therefore, the proposed motion detection technique is used to speed up video event analysis and storage of relevant events.

This CRT scene is described in Fig. 1 and composed of four parts: start line, limit line, tracking area and end line.

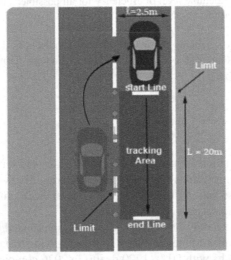

Fig. 1. Car reverse test scene review (bird-eye view) described by [1]

In this research, an appropriate method is proposed to manage the tracking area part of the scene. In this part, the main goal is to successfully detect the car when it comes through the tracking area and extracting the crossed path to further processing. Technically, the algorithm captures the video of the scene after the car crossing the start line event, then, extracts the zone of interest and detects the car exact location and save it to make sure that test rules related to not crossing limit lines nor stopping the vehicle are respected.

The main goal of using the proposed approach for monitoring and supervising driving license exam is to assist the supervisor by offering realistic results which help him to get a better vision and an appropriate judgment of candidate's driving capabilities. The proposed approach in this paper is based on using only one camera which could be installed and simply configured to finally give reliable results with lower coast of energy and chipper equipments compared with other methods (sensors for example).

The paper is organized as follows: in Sect. 2, some related work is presented; and then, an overview of proposed methods is presented in Sect. 3. In Sect. 4, the performance of the proposed mechanism is evaluated using a set of experiments on offline videos. Finally, Sect. 5 concludes the achievements of the paper.

2   Related Works

Detection of moving object is a principal step of many applications such as anomaly detection, human detection, robotic, video surveillance, traffic monitoring, target tracking and others. There are many proposed methodologies of multitudes of researchers on object tracking from videos sequences. In this section, some significant algorithms are presented. Nonetheless, the best method has not yet been developed.

The term of object tracking is used to identify an object position from video sequences as described by authors of [2] and [7]. According to authors of [3], there are three types resembling tracking method: kernel, point and silhouette based tracking.

Point Based Tracking
Point tracking is presented by authors of [8] which is usually used to track vehicles. This method can be probabilistic which is relied on the probability of object movement or deterministic which is based on Histogram of Object Gradien (HOG) introduced by authors of [9].

Kernel Based Tracking
Kernel tracking or also called as appearance based tracking which is relied to the motion of the object from frame to frame. This method can be single view as described by authors of [10] which concerns the case when a single camera is used. And also multi-view which is suitable for multiple cameras as described by authors of [11].

Silhouette Based Tracking
Silhouette tracking is presented by authors of [12], which is used to track objects with complex shapes such as head, hands, and shoulders cannot be accurately defined by geometric shapes.

The kernel based method is the most used when high accuracy is required with relatively less hardware cost. However, the point tracking method consumes less computational cost with reduced accuracy.

The various types of object tracking are illustrated in Fig. 2. Moreover, there are many algorithms for each category; authors of [3] have described those approaches using a comparative study which is tabulated in Table 1.

Table 1. Object tracking methods comparative study introduced by [3]

Category	Method	Algorithm	Accuracy level	Comp. time level
Point tracking	Kalman filter	Kalman filering algorithm	Moderate	Low to moderate
	Particale filter	Recursive bayes filtering	High	Moderate to high
	Multiple hypothesis tracking	MHT algorithm	Low to moderate	Low
Kernel tracking	Simple template matching	Matching region of interest	Low	Low to moderate
	Mean shift method	Expression and location of object and optimal gradient decline	Moderate	Low
	Support vector machine	Positive and negative training values	Moderate	Moderate
	Layering based tracking	Shape representation using intensity	Moderate to high	Moderate
Silhouette tracking	Contour matching	Gradient descent algorithm	Moderate to high	Moderate
	Shape matching	Hough transform	High	High

Finally, it is necessary to mention that the tracking and recognition as completely separate processes and the recognition can relatively need a long time to be applied to the first frame, after which the object may have moved too far to be tracked. This paper managed to get tracking and recognition processes side by side. The already extracted image features are used to calculate homography and then find the model-view matrix. In this way it speeds up the system performance and can work in real time.

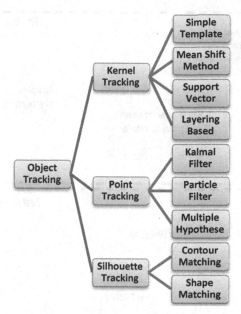

Fig. 2. Object Tracking categories and approaches by authors of [3]

3   Overview of the Method

The main idea is to generate the path made by the vehicle which leads the program to make the best judgment. Therefore, using the generated path, the proposed program could verify his linearity and also calculate the vehicle's speed so the nonstop condition could be verified for the test. Those two criteria are actually evaluated by the supervisors. However, the program would help them to come up with a better judgment using concrete results. The general view of the proposed program of CRT system is described in Fig. 3.

The system could be divided on two principal parts: Pre-processing in which after the frame acquisition from the scene, a Planar Homography Transformation (PHT) is applied on the tracking area in order to get the bird's eye view in order to facilitate the detection process. In the next step, a background subtraction based on Structural Similarity Index (SSI) difference is applied to detect the car exact location. Then, a noise removing using a bilateral filter or edge preserving filter is applied. In that stage, once the car is detected, the post processing stage is started in which the location center of the vehicle's shape will be stored and the car's left and right border will be defined. This makes it easy to detect any possible limit line crossing event. At the end, when the car crosses the end line, the path will be generated for further process.

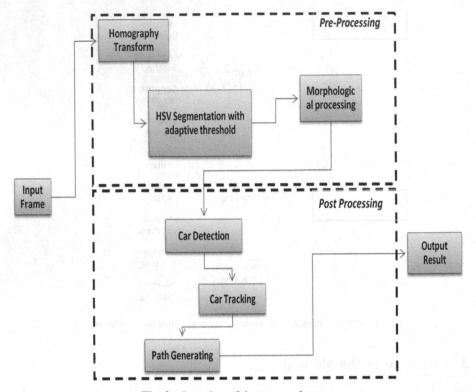

Fig. 3. Overview of the proposed system

3.1 Pre-processing Stage

Homography Transform

The term of homography is defined as the association between world and screen coordinates. Technically, the world 3D coordinates (X, Y, Z) is defined by set of points in screen coordinates (x', y', z'). The camera transformation and homography specifications was discussed by authors of [4] to deduce the overall camera 3 × 4 matrix as follow:

$$\begin{bmatrix} x' \\ y' \\ z' \end{bmatrix} = \begin{bmatrix} m11 & m12 & m13 & m14 \\ m21 & m22 & m23 & m24 \\ m31 & m32 & m33 & 1 \end{bmatrix} \begin{bmatrix} X \\ Y \\ Z \\ 1 \end{bmatrix} \tag{1}$$

According to authors of [4], the Planar Homography is based on the simplification matrix of the camera matrix. The coordinates of point in the 2D plane is defined by (X, Y, 0, 1) where Z = 0. Therefore, since the camera matrix will be multiplied by the homogenous coordinates of the point, the value of m13, m23 and m33 will not be needed because of elimination of Z, the resulted 3 × 3 matrix from Eq. (1) is called Homography Matrix (2).

The Homography matrix is generally used to change the perspective in which the image is seen (to look at it otherwise) and this is made by moving every single point of the image with Homography matrix in order to produce a transformed image with desired vision.

$$\begin{bmatrix} x' \\ y' \\ z' \end{bmatrix} = \begin{bmatrix} h11 & h12 & h13 \\ h21 & h22 & h23 \\ h31 & h32 & 1 \end{bmatrix} \begin{bmatrix} X \\ Y \\ 1 \end{bmatrix} \tag{2}$$

To calculate the value of the homography matrix, a four initial points coordinates must be known of the image (x1, y1; x2, y2; x3, y3; x4, y4) and also the desired coordinates (x'1, y'1; x'2, y'2; x'3, y'3; x'4, y'4). Then the application on the equation above can get the Eq. (3) and (4) as follow:

$$x' = \frac{h11x + h12y + h13}{h31x + h32y + 1} \tag{3}$$

$$y' = \frac{h21x + h22y + h23}{h31x + h32y + 1} \tag{4}$$

In this paper, the homography transform is used to get the bird's eye view using the method described by authors of [5]. According to the same source [5] the transformation process is composed of three main steps: perspective mapping, shearing and enlarging as it is shown in Fig. 4.

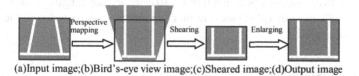

(a)Input image;(b)Bird's-eye view image;(c)Sheared image;(d)Output image

Fig. 4. Bird's eye view process by authors of [5]

HSV Segmentation with Dynamic Thresholding

To extract the exact form of the vehicle crossing the Tracking Area (TA), a segmentation technique based on HSV color space is implemented. The HSV image as it is described in [6] is divided into three components: hue, saturation and value. The hue is the shade of the color to define one color from other; the saturation is the purity of the color and the value represents the brightness of the color.

The idea is to extract and merge the hue and saturation images of the pre-defined region of interest TAi from the current frame i. This method is adopted because the object is possibly exposed to the shadow variation which makes the segmentation more difficult using other known methods. Then, a dynamic threshold value is calculated to extract the foreground of the vehicle.

Morphological Processing

After that, morphological operations (masking, filtering, smoothing) are applied in order to produce a suitable shape of the object for further treatment.

3.2   Post Processing Stage

At the end of pre-processing stage, the shape of the vehicle is detected. Therefore, the next step is to locate the vehicle by detecting the left and right border in order to get the center point which will be saved. The final step is to generate the path made by that car throw the tracking area which is composed by the obtained points.

4   Experimental Result and Discussion

As it is represented in the section above, the first step is to get a virtual bird's eye view which represents a perspective rectification. In the experiment, the bird's eye perspective transformation is performed just after the vehicle crosses the start line. Therefore an HSV segmentation technique is applied in order to get the vehicle's shape after a noise removing process. In that stage, the vehicle's edges will be determined in order to detect wheel's position and also the center point. The process will be repeated for all frames and the obtained center point coordinates X and Y location will be added to the path for further process. This whole process is presented in Flowchart shown on Fig. 5.

In this section, the results of the proposed method will be discussed. The program execution is launched on a desktop computer with Intel Core i5 1.80 GHz CPU, 8 GB RAM. The software is developed on windows 10 using Python language and the OpenCV library. The collection of videos used in this experiment was recorded in the real scene for better evaluation.

4.1   Dataset Description

In order to similate an active scene of exam, a real time videos are recorded by a fixed camera which is installed in front of the tracking area. The dataset used in this research is composed of 25 s videos with 1280 * 720 as resolution. The dataset is composed of 8 video samples divided on three categories: successful test, right limit line crossed and left limit line crossed.

4.2   Proposed Program Execution

In order to have a clearer idea of program's execution, it will be divided on three majors steps: Bird's eye transformation, HSV segmentation, Car's edges and center point detection.

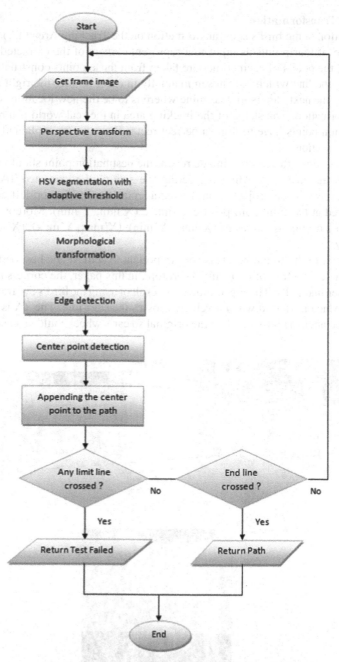

Fig. 5. Flow chart of the proposed tracking method

Bird's Eye Transformation

The application of the bird's eye transformation on the Tracking Area (TA) requires the specification of four points as input that represent corners of the extracted TA. In this experiment, the points of each corner are taken from those points constructing the start line and the end line which are chosen manually in order to have the right angle of the scene. Then, the next step is to determine where is to be the new location of each point after transformation. The shape of the tracking area in the real world is a rectangle, so the destination points have to draw a perfect rectangle as it was explained in Fig. 4 of the previous section.

In order to draw the desired image result, the destination point should be specified manually or automatically while respecting the size of the extracted TA. So all the extracted pixels will be projected on the rectangular result: The top-left corner of TA will be stored at the minimum pixel coordinate (X'min, Y'min). Moreover, the other corners will have pixels values of (X'max, Y'min), (X'max, Y'max), (X'min, Y'max) respectively.

The choice of the initial points is very important to achieve the best performance of the perspective transformation result. Therefore, in this paper, the corners points of the TA are chosen manually. The Fig. 6 shows the result of using the bird's eye transformation on the TA which is started when a vehicle crosses the start line. The TA is represented as near to rectangle as possible than the original so car's wheel could be easily detected.

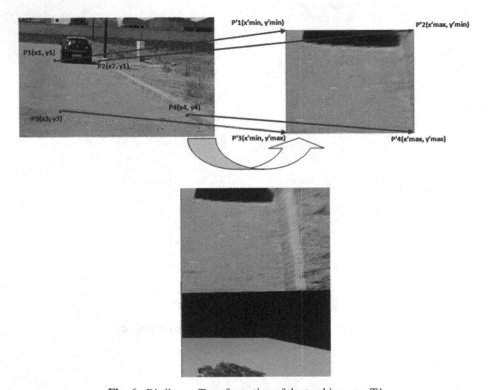

Fig. 6. Bird's eye Transformation of the tracking area TA

HSV Segmentation

The segmentation is a process that aims to separate the object from the background. However, in this paper, the desired object may contain an additional part like the shadow which complicates the treatment and affects the result of the test. Therefore, as it's mentioned before, a segmentation method based on HSV color space will be adopted in which the hue and saturation images are merged in order to get the object's exact shape. Then, the next step is to apply an image binarization using Otsu's thresholding method. The application of the proposed method gives the result illustrated in Fig. 7.

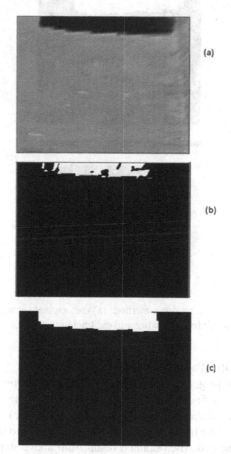

(a)

(b)

(c)

Fig. 7. Result of the proposed HSV segmentation method: (a) original cropped image, (b) H and S merged result with threshold applied and (c) morphological operation result

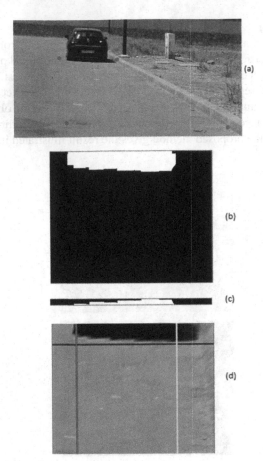

Fig. 8. Result of the proposed tracking method: (a) the original scene, (b) segmentation result, (c) the processed window, (d) lines drawing and car tracking

Car's Edges and Center Point Detection

In this stage, in order to get the location of the vehicle in the tracking area TA, the proposed method is to define three lines: bottom line, left line, right line. Those lines serve to locate the shape of the obtained object shown in Fig. 7(c). The idea is to locate the first white color pixel of the biggest region which will be used to draw the line. However, in order to simplify the task, the obtained result from above step will be cropped by a window of 10 pixels of height starting by the position of bottom line. This process will be repeated for all frames and the points in the middle of left and right lines will be stored. At the end of the test, the stored points will be used to generate the path crossed by the vehicle for further treatment. The Fig. 8 shows the program's result in which the desired object is successfully located in all parts of the tracking area TA.

Program's Performance and Evaluation

The program result or output is represented by the paths crossed by the vehicle through the TA for each test. In this experiment, a path is a collection of the stored center points during the test. Therefore, in order to evaluate the performance of proposed method, we have selected four videos with different scenarios. Table 3 shows the generated path for each case along with the test result. The proposed program successfully generates an accurate result with the ground truth. The generated path is visibly easy to identify, though there is some points which are not fitted with the shape and that is due to effects caused by the environment on the camera (wind, camera shacking...). The Table 2 shows the performance of the proposed program represented by the calculation of recall and precision parameters obtained at the end of each video sequence. The evaluation metrics are defined as follows:

$$recall = \frac{\sum_t TP}{\sum_t N}$$

$$precision = \frac{\sum_t TP}{\sum_t (TP + FP)}$$

$$f - measure = \frac{2 * recall * precision}{recall + precision}$$

Where TP is number of true positives: points are marked in the same way as the vehicle in ground truth, N represents the total marked points and FP is the number of false positives: points are marked outside the way crossed by the vehicle in ground truth.

The evaluation of the test could be objectively done by simple eye observation: analyzing the linearity of the path resulted from the output and also by the variation of speed noticed by the density of points constructing that path. This process will be the subject of our future work in which using a variation of techniques of image processing and classification.

Table 2. The evaluation of the proposed method on video sequences

Video sequence	TP	FP	Recall	Precision	f-measure
Video 1	192	25	0.86	0.88	0.87
Video 2	475	30	0.92	0.94	0.93
Video 3	481	20	0.94	0.96	0.95
Video 4	180	12	0.85	0.93	0.89

Table 3. Exam test results and generated path using the proposed method

Video sequence	Path	Path type	Result
Video 1		Straight	Passed
Video 2		Zig-zag	Passed
Video 3		Left oriented	Failed: left line crossed
Video 4		Right oriented	Failed: right line crossed

5   Conclusion and Perspectives

In this paper, another part of CRT supervising system is presented. In which, we proposed a real-time tracking method based on HSV and perspective transform in order to characterize the vehicle in the tracking area in order to automatically supervise the car reverse test of driving license exam. To this stage, the program uses offline video sequences of real scenes as input. The process is started when the vehicle crosses the start line. Then, a perspective transform is applied on the tracking area to obtain the bird's eye view. After that, a segmentation method based on HSV color space is used in order to get the exact shape of the subject and define his edges. The program successfully detects the left and right crossing events to finally determine whether the candidate is passed or failed the exam. At the end, the program generates the path followed by the vehicle through the tracking area. The experimental result shows that our approach achieves reasonable outputs comparing with the ground truth. The next stage in our research is to process the obtained path as input in which the program will try to unify it and analyze it to finally evaluate subjectively the candidate's performance using a classification technique.

References

1. Azi, A., Salhi, A., Jourhmane, M.: Line area monitoring using structural similarity index. Int. J. Adv. Comput. Sci. Appl. **10** (2019)
2. Balasubramanian, A., Kamate, S., Yilmazer, N.: Utilization of robust video processing techniques to aid efficient object detection and tracking. Procedia Comput. Sci **36**, 579–586 (2014)
3. Nagar, R.R., Diwanji, H.M.: A review of RealTime object detection and tracking. Int. J. Sci. Res. Sci. Eng. Technol. (IJSRSET) **3**(6), 598–603 (2017)
4. Sihombing, D.P., Nugroho, H.A., Wibirama, S.: Perspective rectification in vehicle number plate recognition using 2D-2D transformation of Planar Homography. In: 2015 International Conference on Science in Information Technology (ICSITech), pp. 237–240 (2015)
5. Luo, L.-B., Koh, I.-S., Min, K.Y., Wang, J., Chong, J.: Low-cost implementation of bird's-eye view system for camera-on-vehicle, pp. 311–312 (2010)
6. Maheswari, S., Korah, R.: Review on image segmentation based on color space and its hybrid. In: 2016 International Conference on Control, Instrumentation, Communication and Computational Technologies (ICCICCT), pp. 639–641 (2016)
7. Tiwari, M., Singhai, R.: A review of detection and tracking of object from image and video sequences. Int. J. Comput. Intell. Res. **13**(5), 745–765 (2017)
8. Yilmaz, A., Javed, O., Shah, M.: Object tracking: a survey. ACM Comput. Surveill. **38**(4), Article 13 (2006)
9. Bilinski, P., Bremond, F., Kaaniche, M.: Multiple objects tracking with occlusions using HOG descriptors and multi resolution images. In: The International Conference on Imaging for Crime Detection and Prevention (ICDP), London, UK (2009)
10. Chau, D.P., Bremond, F., Thonnat, M.: A multi feature tracking algorithm enabling adaptation to context variations. In: The International Conference on Imaging for Crime Detection and Prevention (ICDP), London, UK (2011)
11. Colombo, A., Orwell, J., Velastin, S.: Color constancy techniques for ReRecognition of pedestrians from multiple surveillance cameras. In: Workshop on Multicamera and Multi-modal Sensor Fusion Algorithms and Applications (M2SFA2) (2008)
12. Angadi, S., Nandyal, S.: A review on object detection and tracking in video surveillance. Int. J. Adv. Res. Eng. Technol. **11**(9), 1033–1042 (2020)

Author Index

Printed in the United States
by Baker & Taylor Publisher Services